Because of the dynamic nature of the Internet, any web addresses or links contained in this book may have changed since publication and may no longer be valid. The views expressed in this work are solely those of the author and do not necessarily reflect the views of the publisher, and the publisher hereby disclaims any responsibility for them.

This book is a work of non-fiction. Unless otherwise noted, the author and the publisher make no explicit guarantees as to the accuracy of the information contained in this book and in some cases, names of people and places have been altered to protect their privacy.

Any people depicted in stock imagery provided by Getty Images are models, and such images are being used for illustrative purposes only. Certain stock imagery © Getty Images.

Lulu Publishing Services rev. date: 06/14/2018

Comprehensive Reviews

Parts V and VI:
On
Probability of God
and
Proof of Heaven

Arieh Ben-Naim

This book is dedicated to all those who read or intend to read either
Unwin's book: Probability of God
or Alexander's book: Proof of Heaven

פֶּתִי, יַאֲמִין לְכָל-דָּבָר; וְעָרוּם, יָבִין לַאֲשֻׁרוֹ.
נָחֲלוּ פְתָאיִם אִוֶּלֶת; וַעֲרוּמִים, יַכְתִּרוּ דָעַת.
משלי, יד; טו,יח

Preface to the series of books: Comprehensive Reviews.

Since I retired I have written several popular science books, as well as read many books which were written by other authors. Some books were not only highly entertaining, but quite informative as well, allowing me to explore domains other than my own. However, there are two sides to every coin. I read quite a few books which were a total let-down, and more seriously, were quite misleading and deceiving.

It behooves me to write these series of books, and present to the readers critical reviews of all those misleading creations, which unbeknownst to the authors might have done damage, sowed disinformation instead of providing valid and correct information.

Indeed, I already wrote one such book entitled: "The Four Laws that do not Drive the Universe," Ben-Naim (2017). The book's concept touches on the Four Laws of Thermodynamics. The particular choice of the title of my book is by no means accidental, as it was meant to refute a book written by Atkins, entitled "Four Laws that Drive the Universe." My book is not merely a critical review, but more importantly, it presents a new perspective of the four laws, particularly the Second Law of Thermodynamics.

I strongly oppose Atkins' views about the Second Law because I believe that they are wrong and misleading. In addition, claiming that the four laws "drive the universe" is totally unfounded.

In these series of critiques entitled "Comprehensive Reviews," I will focus mainly on criticizing some books which I believe propagate erroneous ideas. Occasionally I will present my own views, or what I believe are the correct views.

These books are addressed to the readers who either have read or intend to read the books that I have criticized.

I am well aware of the risk I am taking by undertaking this project. However, I strongly believe that it is my duty to undertake this unpleasant task. I hope not only to be able to point out to the readers where the authors went wrong, but to educate them as well on how to be more critical whenever they read books which do not necessarily belong to their field.

When I ran out of words in describing the merits (or lack of merits) of some ideas, I resorted to using a Hebrew word *kishkush* and its plural form *kishkushim,* which means scribbles, or better yet, meaningless scribblings. The Hebrew word not only *means* nonsense; it also *sounds* nonsense. Therefore, I occasionally resorted to using this word so as not to bore the readers with the same adjectives such as nonsense, meaningless, and silly. Some typical kishkushim are given after the list of abbreviations.

It is my intention in this series of books to point out some serious flaws appearing in some popular science books. I am well aware that mistakes are done in science, and my writings are not exceptions. I also hope that no personal offence will be taken by anyone whose work I have criticized. I made a great effort in criticizing the content of the book, and not the author himself.

As always, I encourage the readers of my books to write to me, and let me know of any comment or criticism on my books.

Arieh Ben-Naim

Department of Physical Chemistry
The Hebrew University of Jerusalem
Jerusalem, Israel
Email: ariehbook@gmail.com
URL: ariehbennaim.com

Acknowledgements

I am grateful to Steven Bottomley, Robert Engel, Robert Goldberg, Jose Angel Sordo Gonzalo, Douglas Hemmick, Shannon Hunter, Richard Henchman, Maik Jacob, and Mike Rainbolt for reading parts or all of the manuscript and offering useful comments.

As always, I am very grateful for the graceful help I got from my wife, Ruby, and for her unwavering involvement in every stage of the writing, typing, editing, re-editing, and polishing of the book.

Examples of Kishkushim:

- **Entropy is a measure of disorder**
- **Entropy is the cause of everything that happens**
- **Entropy creates life**
- **Entropy, together with time ravages everything**
- **Entropy explains Time**
- **Life is a delicate dance between "Entropy and Entropia"**

Part V: Review of:

The Probability of God:

A simple calculation that proves the Ultimate Truth

By: Stephen D. Unwin, PhD

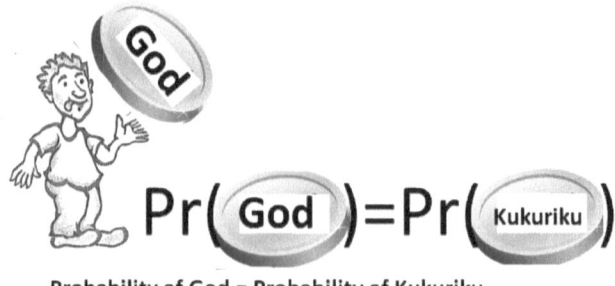

Probability of God = Probability of Kukuriku

Preface to parts V and VI

The two books reviewed in the following pages of this book have a common denominator; they are written by scientists using scientific language (or perhaps unwittingly, abusing scientific language) on a non-scientific subject. Perhaps, a more apt description of the common denominator is that the book deals with two fictitious subjects; God and heaven, respectively.

Lest believers in God and heaven, or in both, take my criticism on a negative vein, let me say that nowhere in my review do I criticize their beliefs, or the author's. I totally respect the beliefs of everyone, including the authors and the readers of these books. My criticism is based solely on the *methods* used by the authors in order to prove their point, and which I will show to be misleading, if not downright deceitful.

There is a difference in the extent of misinformation and disinformation provided by these books. First, Stephen Unwin has a PhD in theoretical physics. In his book he abuses probability theory, specifically Bayes' theorem in order to prove something which is not even defined. This effectively renders the entire book deceptive. The second author, Alexander Eben is a neurologist by profession. Unlike the other author, he did not misuse, or abuse any theorem, and did not prove anything. However, by using the word "proof," in the title, he violated the ethics of scientific writing. While he does not employ any mathematical theorem to prove his point, he lends credibility to his NDE (near-death experience) as a *proof* of heaven, by capitalizing on his scientific background, and makes use of some scientific, neurological terms plucked from neurology.

1. The title: The Probability of God

In the following pages, I will criticize in more details why the usage of probability in this book is inappropriate, and misleading, thus rendering the title, meaningless. There is no probability of a chair, nor a probability of an atom, or a probability of Stephen Unwin, and certainly no "Probability of God." More specifically, a probability, in the theory of probability, is assigned to an event; the probability that the result "4" in throwing a die is 1/6. In logic, one also uses the probability for a proposition. We say that the probability that this book was written by Shakespeare is 0.1 or 10%. But there is no probability of an *object*, and certainly no probability of something that is not even defined. Therefore, the title of this book is meaningless, as much as the title, "The probability of Kukuriku," is meaningless.

2. The subtitle: A simple calculation that proves the Ultimate Truth

This subtitle is not only deceptive, and misleading, it is downright untrue. First, it is not a simple calculation. What the author does is a far more complicated calculation than when one usually does when using Bayes' theorem. Second, the author does not prove anything, but simply *abuses* the terms "probability" and "conditional probability," and then abuses Bayes' theorem to "prove" his point. As I will show in details, using his methods, I could prove anything I want. As I will demonstrate by using the author's methods, I could prove for example that my dooglag's flinogler is true. After reading this book, the reader can use the same method to prove just about anything they want – suggestions by the reader are welcome.

Finally, the "Ultimate Truth" does not need a proof. By using the term "Ultimate Truth" the author already made a statement about the truthfulness of his result, so why the need for a proof?

By this very subtitle the author betrays himself and discredits his message in this book.

3. Chapter 1: Modest Objectives

The book opens with the following fancy sentence:

"Do you realize there is some probability that before you complete this sentence, you will be hoofed insensible by a wayward, miniature Mediterranean ass?"

Clearly, the author presumes that the reader knows what probability means. In this particular example, the author uses the term probability in its colloquial sense.

Then, surprisingly the author tells the reader, who had apparently survived a hoofing incident, that he was saved by an aspect of probability.

I have taught probability theory for many years, and I have no idea which "aspect of probability" has saved the reader. This is a typical obscure, nearly meaningless sentence which is enhanced by telling us that:

"Probability is the subject of this book:

God, faith, and probability"

This statement not only deludes the reader but also exposes the author's ignorance of the "subject of probability." First, the subject of the book *is not probability*. Second, probability theory does not deal with either "God," or with "faith." What remains is that Probability deals with probability – this is tautology – yet, the book is not about probability theory.

The first meaningful, yet outright misleading statement comes next when the author assures us that he will use the term "probability" in its "strict mathematical sense," and not in the fuzzy and ambiguous way it is used in common language.

If you haven't read the book yet, you will expect to read about *probability* in its "strict mathematical sense." Unfortunately, you will find out that the author failed to deliver on his promise. because the book does *not* discuss probability, "in its strict mathematical sense," but rather in the *most extreme*, fuzzy and ambiguous sense. From the very first page, as well as in the entire book, you will not find any "probability in its strict mathematical sense. Therefore, the above quoted sentence is an outright deception. To clarify what I have written above consider the following probabilities:

1. The probability that the outcome "Head" will result in tossing a fair coin is ½, or 50%.

2. The probability that the outcome "4" will result in throwing a fair die is 1/6.

3. The probability that tomorrow morning it will rain in Jerusalem is 30%

4. The probability that the sun will rise tomorrow is 99.99%.

5. The probability that this book was written by Shakespeare is 0.01%

6. The probability that Kukuriku will read and enjoy this book is 75%.

All these probabilities have some meaning as an extent of our belief in some event. However, the first two examples are, what may be referred to as "scientific" probabilities. The third and the fourth are based on accumulated statistics and some other reasoning. The fifth is a pure guess, or, if you like, you can call it a pure non-scientific probability. The sixth is not even "non-scientific," it is a meaningless statement. The reader should be aware of the fact that the term "probability" as used by Unwin, is only in the last sense!

For the readers of the book who are not familiar with the concept of probability, I have added note 1, in which a very brief discussion of probability is presented. Also, in note 2, I present Bayes' theorem which was unfortunately abused by the author of the book "Probability of God."

As I mentioned before, the author presumes that the reader knows what probability means, because nowhere in the book does he define, or at the very least, describe the term probability. Instead, we are presented with a very trivial statement about probability, namely that probability is not just any number.

All probabilities, be they mathematical, or "fuzzy and ambiguous," are expressed in *numbers,* not by letters, not by cows, and not by chickens. Indeed, probability is not "just any number," but very specific numbers such as $\frac{1}{2}$ and $\frac{1}{6}$ in the first two examples above. Ironically, all the numbers given in this book can be said to be arbitrary, fictitious or simply "any number" that comes into your mind.

Next, the author explains the word *faith* in its religious sense; faith in God. This is clearly superfluous; faith does not need to be defined or explained, you either have, or do not have faith in God.

Finally, the author tells us that his objective in writing the book is to calculate the numerical probability that God exists. Then, he will discuss the relationship between this probability and the notion of religious faith.

This objective failed miserably. The author *does not* calculate the numerical probability (in the sense he explained in the first page) that God exists. Instead, he calculates a fictitious, and meaningless number which he refers to as the probability that "God exists." Also, he does not determine any relationship between the mathematical probability, and the notion of religious faith.

Having explained his objectives, the author warns us that he does not plan to calculate the probability of God to an accuracy of, say, four significant digits such as 81.91%. That would be absurd! This is true. As in the third and the fourth examples provided above, the extent of one's belief cannot be stated with any accuracy. However, the author tells us that he is modest and he will be satisfied with an accuracy of two significant digits.

What he actually does is the calculation of something, he calls the probability that "God exists," with *zero* significant digits, which is tantamount to a completely arbitrary number – based on arbitrary "logical" deductions, seemingly based on Bayes' theorem. Again, by this very sentence the author betrays himself.

After presenting some examples of fuzzy probabilities, the author adds another claim that I am not sure I understand, that having "an accurate" of the probability that God exists is a critical step in any religious decision making.

Anyone who faces "day-to-day religious decision making," does not need either an accurate, or an inaccurate probability that God exists. In fact, one does not need any such probability, accurate or approximate, to make any religious decision. Anyone who knows elementary probability theory, must know that an "accurate probability that God exists" *does not exist!*

Then, the author tells us that Greek philosophers abandoned the "notion that God could be understood within logical, mathematical or even philosophical framework. Instead, the greatest minds in history "have produced incomprehensible treatises couched in impenetrable vocabulary, leading to vacuous conclusions." This is exactly what the author produces in this entire book.

The author further explains that all these great minds in history failed because they formulate the question in a strictly binary, deterministic way. They asked:

"Is there a God, yes or no?"

Whereas the author will tackle this problem from a different perspective, which is to calculate the probability that the "true answer" is yes; that God exists.

I think that the author does not really understand probability theory. To the question, "Is there a God?" there are three possible answers. "Yes,"

"No," or "I don't know." On the other question, "What is the probability that God exists?" there could be an infinite number of answers, assuming one understands the question.

Then, he concludes that he believes that his approach to the question is the one of intellectual humility.

I do not think that this approach is one of "intellectual humility." There is simply no way to assign probabilities to questions involving faith.

In the next few pages, the author briefly discusses his background in theoretical physics, and his work on *quantum gravity*. This background is provided for two reasons: First, because modern physics has embraced probability theory, which was absent in "classical" physics; second, to impress the reader that he, the author, comes from a scientific background and not from a religious background.

On page 7, the author explicitly reveals the reason for telling us his scientific background, more specifically his "probability qualifications," on one hand, and his "God qualifications."

The next few pages describe his "God qualifications" background which, combined with his scientific background qualifies him to calculate the probability that God exists. As we shall see, neither of his qualifications, qualify him to calculate the "probability that God exists."

The chapter is concluded with the promise to define the terms: *God*, *faith*, and *probability* in the next chapter.

If you read the next chapter, as well as the entire book you will find no definition of *God*, no definition of *faith*, and no definition of *probability*.

The truth is that none of these are definable concepts. As for the title of the chapter, the "modest objective," is misleading. In fact, it is an over ambitious objective, which to the best of my knowledge, no one has ever undertaken before. And as we shall soon see the author achieves nothing!

4. Chapter 2: Not just any God

The chapter starts with a "Question for the reader:"

"What shall we mean by the word God in our central proposition: God exists?"

As a reader of Unwin's book, I have no idea what is meant by God; either before, or after reading his book. Readers who know what the author means by the word God, do not need any estimates of the probability that God exists.

The objective of the book is repeated on page 14, namely to calculate probability that the proposition "God exists" is true. Then he denotes by "G" the proposition "God Exists."

Note that the "probability of Proposition G," is different from the meaningless title of the book "Probability of God."

Then, the author correctly emphasizes that a probability cannot be attached to a meaningless proposition, and provides an example, the proposition:

I did not have sexual relations with that woman.

The author correctly points out that such a proposition is meaningless without defining what is meant by "sexual relations."

Here is a hidden pitfall in his correct assertion that one needs to clearly define the proposition before calculating any probability. He focused his attention in defining "sexual relations," presuming that everyone already knows who "that woman" is.

The author claims that propositions such as "God is the universe," or "God is love," will be assigned probabilities of 100%. Any probability assignment less than 100% would mean that one doubts the very existence of the "Universe" or "Love." This kind of logic is misleading. As in the case of "that woman," which the author unjustifiably assumes that everyone already knows, it is also not true that the proposition "God is the Universe," or "God is Love," are meaningful without defining who *God* is. To this, the author also adds two famous quotations from Albert Einstein: "God does not play dice," and "God is subtle but he is not malicious." I do not know exactly why the author brought these two quotations here. Perhaps to lend credibility to his calculations or to convince the reader that when he talks about the Proposition "God Exists," he knows what he is talking about. I am not sure whether Einstein believed or did not believe in the existence of God. What I am certain of, is that Einstein would not have accepted any calculation of the "probability of God," and certainly not the calculations presented in this book.

The author does not define God. In the rest of this chapter he quotes several, well-known statements made by philosophers and physicists, from Spinoza, to Einstein, to Weyl.

On page 19, the author says that his analysis is *not* about the "God of Spinoza, Einstein, or any other fancy philosophical school," but instead he will refer to the

"God of Christians, the Jehovah of Jews, the Allah of Muslims…, et cetera."

The fact is that neither, the "God of Spinoza, or Einstein," (which is *not* the subject of his analysis), nor the "Person-God of major faiths," (which is the subject of the analysis) are well-defined. Thus, this apparent "clarification" about the God in the proposition G, actually blurs the meaning of his central proposition. It is far clear that the God of Christians, the Jehovah of the Jews, and Allah of Muslims is the same God. Perhaps, the author should have written three books on these three different Gods!

The chapter which intends to define God in the proposition G: "God exists" ends up with a vague conclusion that he will use the word "God" in the "traditional sense" associated with the monotheistic faiths, and not in a philosophical, abstract sense.

It goes without saying that if you believe in God "in the traditional sense," you do not need a probability that "God exists." If you however, do not believe in the "traditional sense" of God, then you are left with an undefined proposition; you do not know what God means, and what it means that God exists. You are left baffled like all the other readers who do not know what "sexual relations" mean, nor "who that woman" is. Thus, in effect the book is addressed to readers who already know who God is, "in the traditional sense," and those readers do not need any estimate of the probability that God exists!

5. Chapter 3: You are here

This chapter deals with an inane question to which an equally inane answer is provided. The question of this chapter is:

"Do the general cosmology, physics, and biology of our universe constitute relevant evidence in assessing the probability that Proposition G is true?"

This is actually a meaningless question like asking; "Does physics of our universe constitute relevant evidence in assessing the probability of "proposition X is true?" Clearly, one cannot answer such a question without knowing what X means.

Although neither Entropy, nor the Second Law is mentioned explicitly they are implicitly referred to when the author says that systems "enjoy chaos" far more than "order."

Perhaps the author "knows" that some, or all systems enjoy chaos far more than order," but I do not know that, and I doubt very much that this is true. Of course, I know that many authors of popular science books write about the idea of systems that tend to chaos rather than to order. This is an untrue statement, and it follows from misunderstanding the Second Law of Thermodynamics.

"We would never expect, for example…that my son's bedroom to become tidier without the imposition of some miraculous force."

The last statement clearly alludes to the erroneous view of the Second Law of Thermodynamics (for details, see Ben-Naim (2016b,

2017a,2017c,2018c). Besides, I do not see the relevance of these incorrect statements to the subject of the book.

From these considerations, the author continues with the *anthropic principle* which claims that the constants of physics such as the speed of light, and Planck constant are fine-tuned so that life could evolve in the universe. I do not see any relevance of all of these to the book's main subject. However, I fully concur with the author as he says on page 33 that he believes that any science-based argument for or against the existence of the Person-God is troublesome.

As well as his final conclusion:

The physical and biological laws of our universe and the phenomena to which they give rise do not provide meaningful evidence either for or against the proposition that God exists.

True! And that should be enough reason not to write this book!

6. Chapter 4: The good Reverend Bayes

This is the book's central chapter where Bayes' theorem is presented before applying it to calculate the probability of the proposition G.

The chapter starts yet again with the concept of probability and stressing once more that he will use the term probability in its *mathematical sense*, and not in "some every day, vague, qualitative sense." As we shall see, in this and in the following chapters, the author uses *nowhere*, probabilities in their strict *mathematic sense*, but at best, he uses the vague and qualitative sense, and at worse, in a totally meaningless sense.

The author's trick to achieve his goals is by making use of Bayes' theorem, which is a widely accepted, respectable mathematical theorem involving *conditional probabilities*. The author latches onto the Bayes' theorem with the aim of lending credibility to his calculations. However, the probabilities which are fed into the theorem are fictitious, arbitrary numbers which inevitably lead to fictitious arbitrary results.

The central concept in Bayes' theorem is the *conditional probability*. As in the case of probability discussed in Note 1, the conditional probability can either be mathematical, qualitative or vague.

Consider the following proposition:

G = "This book was written by Shakespeare"

One can assign a probability of 10% to this proposition. This number means that the extent of your belief in Proposition G is true. Clearly, different people will assign different probabilities according to their extent of belief in this proposition. There is nothing of a mathematical sense in this probability. Suppose, instead of proposition G, I suggest proposition G^*

$$G^* = \text{"This book was written by Kukuriku"}$$

Suppose I tell you that the probability of G* is 10%, you will feel that this number is outright meaningless as long as you are not sure who, or what Kukuriku is. The two propositions G and G^* look similar, and in fact they are similar to someone who has never heard of Shakespeare, and who has no idea who Kukuriku is. The two propositions would be different if one knew who Shakespeare was, but had no clue as to who Kukuriku is.

Let us assume that you know who Shakespeare is, and I tell you that this book was printed in 2017. Let us denote this *new information* by NI,

$$NI = This\ book\ was\ printed\ in\ 2017$$

Clearly, having the new information (NI) would affect your extent of belief in proposition G, if we denote by P(G) the prior probability of G, and by P(G|NI) the posterior probability of G given the new information NI. Clearly, (G|NI) < P(G), i.e. the posterior probability is smaller than the prior probability.

In probability theory, the posterior probability (G|NI) is referred to as *conditional probability*. As a simple example, suppose I tell you that I threw a die, and I asked you for the probability that the outcome was "4." If I also tell you that the die is "fair," then your assigned probability to the event "4" will be 1/6, or 100/6%. As I discussed in Note 1, this probability is also a measure of the extent of your belief. However, everyone who uses probability theory would accept the assignment of probability 1/6 to the event "4."

Next, I tell you some new information; that the result is *even*. What is the conditional probability of the result "4," *given* the additional information that the result is *even*? It does not need much thinking to estimate that the modified or the conditional probability of the event "4," given the new information "even" is

$$P["4"|"even"] = \frac{1}{3}$$

which is larger than the prior probability

$$P["4"|"even"] > P["4"]$$

It is easy to find new information which decreases the probability of "4." For instance, if I tell you that the result is "odd" then the conditional probability

$$P["4"|"odd"] = 0 < \frac{1}{6}$$

On page 41, the author explains that the Bayesian probabilities is an expression of a degree of belief. In fact, any probability has some degree of built-in belief. The difference between an arbitrary probability, and a mathematical probability (as well as conditional probability) is that in the mathematical probability everyone who uses these probabilities agrees with their numerical values. This is true for both the classical and the frequency "definitions" of probability – see Note 1.

In Note 2 we derived the Bayes theorem. Let us discuss a simple example.

We are given a board as shown in figure 1. I throw a dart on a board A. The board is divided into two regions A_1 and A_2, such that the union of the two regions is equal to the entire region of the board, write this as $A = A_1 \cup A_2$.

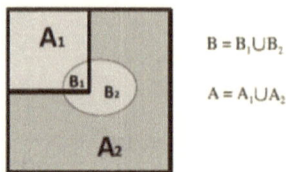

Figure 1. A board with two regions

I tell you that the dart hit the board, and that all the points on the board are equivalent. We shall not be interested in the exact mathematical *point* at which the dart hit the board. Also, we neglect the possibility that the dart hit the line dividing between the two regions A_1 and A_2. Figure 1.

What is the probability that the dart hit the region A_1?

We assume that the probability of hitting any region on the board is simply the fraction of that area from the total area of the board. If the total area of the board is a, and the areas of A_1 and A_2 are $a_1 = a/4$ and $a_2 = 3a/4$ respectively, we assume that the probability of the "event" A_1 (i.e. hitting the region A_1) is

$$P(A_1) = \frac{a_1}{a} = \frac{1}{4}$$

and likewise, the probability of the "event" A_2 is

$$P(A_2) = \frac{a_2}{a} = \frac{3}{4}$$

Note carefully that these probabilities are *assumed* to be correct. There is no proof of those contentions. One can suggest to play this game many times and find out that about ¼ of the times the dart will hit A, and ¾ of the times hit A_2, but this is not the proof of the contention.

Now that we know the probabilities of the two events A_1 and A_2, let us introduce a modification of the problem. Suppose that I have painted with blue some region on the board denoted B in the Figure 1. You are told that 10% of region A_1 was painted and 20% of region A_2 was painted.

What is the percentage of the total area of the board that was painted? If you are in a rush to answer this question you might be tempted to answer $10\% + 20\% = 30\%$ which is wrong. Let us calculate:

The area of A_1 is a_1. Therefore, the painted area of A_1 is $\frac{10}{100}a_1$, and the painted area of A_2 is $\frac{20}{100}a_2$. The *total* painted *area* is therefore

$$b = \frac{10}{100}a_1 + \frac{20}{100}a_2$$

The fraction of the total board painted in blue is also the probability of hitting the region B.

Therefore,

$$P(B) = \frac{b}{a} = \frac{\frac{10}{100}a_1 + \frac{20}{100}a_2}{a}$$

$$= \frac{10}{100}P(A_1) + \frac{20}{100}P(A_2)$$

$$= \frac{1}{10} \times \frac{1}{4} + \frac{2}{10} \times \frac{3}{4} = \frac{7}{40}$$

i.e. the total painted area is 17.5% of the board, quite different from the quick (and erroneous) estimate we did by summing 10% and 20%.

Next, we ask what is the conditional probability of hitting the painted area B, *given* that the dart hit A_1? Given that A_1 occurred, the probability of hitting the painted area is simply the fraction of A_1 which is painted.

$$P(B|A_1) = \frac{1}{10}$$

Again, you can derive this result from the classical definition of probability.[1] The total area of A_1 is a_1, and the painted area in A_1 is $\frac{10}{100}a_1$, therefore the fraction of the painted area in A_1 is $\frac{10}{100} = \frac{1}{10}$.

Likewise, we have the probability of hitting the painted area in A_2

$$P(B|A_2) = \frac{2}{10}$$

At this point, we have all the required probabilities:

$$P(A_1) = \frac{a_1}{a}, \quad P(A_2) = \frac{a_2}{a} \qquad (1)$$

$$P(B|A_1) = \frac{1}{10}, \quad P(B|A_2) = \frac{2}{10} \qquad (2)$$

These are sometimes referred to as the *a priori* probabilities, i.e. these are given in advance, and later we shall calculate *a posteriori* probabilities.

The Bayes' theorem is concerned with calculating a kind of "inverse" probabilities. You are given the probabilities as in (1) and (2), and you are asked about the *inverse conditional* probabilities, i.e. given that the dart hit the painted area B, what is the probability that it is in A_1? In other words, we want to find the conditional probabilities $P(A_1|B)$ and $P(A_2|B)$. You can see that in the latter probabilities the roles of A_1 and B (and of A_2 and B) are "reversed" compared with the role in (2). In other words, we reverse the roles of the *condition* and the outcome events.

To calculate $P(A_1|B)$ we use the definition of the conditional probability (see Note 1)

$$P(A_1|B) = \frac{P(A_1 \cap B)}{P(B)} \qquad\qquad (3)$$

Remember $P(A_1|B)$ is the probability of A_1 given B. $P(A_1 \cap B)$ is the probability that A_1 *and* B occurred.

Next, we want to rewrite the right-hand side of the equation (3) in terms of the quantities we already know. Using again the definition of the conditional probability, we have:

$$P(A_1 \cap B) = P(A_1)P(B|A_1) = \frac{a_1}{a} \times \frac{1}{10} = \frac{1}{40} \qquad\qquad (4)$$

Note carefully that the quantity $P(B|A_1)$ is the probability of B *given* A_1. Furthermore, we already calculated the probability of hitting B, which we can rewrite as:

$$P(B) = P(B \cap A_1) + P(B \cap A_2)$$

$$= P(B|A_1)P(A_1) + P(B|A_2)P(A_2)$$

$$= \frac{1}{10}\frac{a_1}{a} + \frac{2}{10}\frac{a_2}{a} = \frac{1}{10} \times \frac{1}{4} + \frac{2}{10} \times \frac{3}{4} = \frac{7}{40} \qquad\qquad (5)$$

The first line of the equation (5) means that the region B is simply the sum of the areas that B "cuts out of A_1" and the area that B "cuts out of A_2." See the two areas B_1 and B_2 in Figure 1. The equality (5) is sometimes referred to as the *total probability theorem*. As you can see this theorem is nothing but a statement that the area of B in Figure 1 is the sum of the areas B_1 and B_2. Next, we wrote each of the joint probabilities in terms of the conditional probabilities, and in the final step we plugged-in the numbers we already have. The result we need is therefore

$$P(A_1|B) = \frac{\frac{1}{40}}{\frac{7}{40}} = \frac{1}{7} \qquad (6)$$

If you followed me so far in calculating the required probability, you have just *derived* the Bayes' theorem. There is nothing more in Bayes' theorem than what you have already done. It is simply rewriting the conditional probability in a slightly more complicated, albeit more useful way.

Consider the following probabilities:

1. The probability of the outcome "4" in throwing a fair die is 1/6.
2. The probability that this book was written by Shakespeare is 0.1 (or 10%).
3. The probability that this book was written by Shultingford is 0.6 (or 60%).

You should feel that these probabilities have different *"statuses,"* "meanings," or *"significance."* The first probability seems correct even if you do not know mathematics. The reason is that you can "verify" or "prove" that the probability is 1/6 by throwing a die many times, and you will find that about one-sixth of the results will be "4." There is an element of *belief* in this proof, however everyone accepts it assuming he/she understands what a fair die is.

Not everyone will accept (2). You should "know" that you cannot prove it, nor verify it by doing a series of experiments. Finally, you must "feel" that the third probability (3) is arbitrary, not because you cannot verify it, but simply because you do not know who Shultingford is, and whether such a person even exists.

Let us summarize by saying that 1/6 in case (1) measures the extent of our belief that the result is "4." The number 0.1 is a measure of the extent of belief that whoever wrote (2), thinks that Shakespeare wrote this book. Also, the value of 0.6 is a measure of the extent of belief that whoever wrote (3), thinks that Shultingford wrote this book.

As you correctly feel the status of these three probabilities are different. Let us try to quantify this feeling by viewing the three statements (1), (2), and (3) as propositions. Then, I ask you:

(i)   What is the extent of your belief that proposition (1) is true?

(ii)   What is the extent of your belief that proposition (2) is true?

(iii)   What is the extent of your belief that proposition (3) is true?

The answers are: (i), your answer should: 1 or 100%; (ii), you should answer that you do not know, or say 50%, or any other number depending upon *your belief* that whoever wrote (2) knows what he/she is talking about; (iii), you should also answer "I do not know," or if you feel that whoever wrote (3) has no idea what he/she wrote, you would say, zero.

You see that the answers to questions (i), (ii) and (iii) are very different from the probabilities stated in (1), (2) and (3), respectively.

As we have seen one of the "ingredients" inputted in the Bayes' Theorem is the Total Probability Theorem.

In case of an experiment with two outcomes, e.g. the case of Figure 1 we wrote

$$B = A_1 \cap B + A_2 \cap B \qquad (7)$$

and the corresponding probability (for sums of disjoint events)

$$P(B) = P(A_1 \cap B) + P(A_2 \cap B)$$

$$= P(B|A_1)P(A_1) + P(B|A_2)P(A_2) \qquad (8)$$

For an experiment with n outcomes (again assuming that the outcomes are disjoint and their sum is equal to the sample space Ω), we have:

$$P(B) = \sum_{i=1}^{n} P(B|A_1)P(A_1) \qquad (9)$$

The question we want to discuss now is whether we can apply Bayes' Theorem for the case of a proposition.

More specifically, we discuss the proposition G

$$G = God\ exists \qquad (10)$$

And suppose we have new information or new evidence, and we want to estimate the modified probability of G given the evidence E.

Unwin writes this modified probability as:

$$P(G|E) = \frac{P(E|G)P(G)}{P(E|G)P(G)+P(E|G^*)P(G^*)} \qquad (11)$$

Remember that the numerator $P(E \cap G)$, means the probability that both E and G are true. We shall discuss specific evidence, or what the author calls "Evidentiary Area," let us choose the proposition E as

$$E = The\ existence\ of\ religious\ experiences$$

Here are the main difficulties in the applicability of Bayes' Theorem for this case:

1. We do not have any idea what $P(G)$ is.

2. We do not have any idea what $P(E \cap G)$ is.

3. We do not know over which events we must sum to obtain the Total probability of E, see equations (9) and (11).

On pages 44-45, Unwin correctly points out that equation (9) is written on the assumption that there are only two possibilities; either G is true, or false. In fact, there is another possibility which is mentioned on page 45:

Theists would say: G is true and G^* is false.

Atheists would say: G is false and G^* is true.

Agnostics would say: Don't know whether it is G or G^* is that is true.

Here, he mentions the third possibility, but in using Bayes' theorem he uses the first two probabilities only, when there is in fact, a fourth possibility:

I do not know what God is. Therefore, I cannot make any statement regarding either G or G*

Recall that the author carefully pointed out that proposition: "I did not have sexual relations with that woman," is meaningless unless one defines clearly what "sexual relations." This is correct, but he overlooked the possibility that the reader might not know who "that woman" is, rendering the whole proposition meaningless!

The same comment applies here. The proposition G presumes that we know who, or what God is, and that there are only two possibilities; either God exists, or not. This very assumption already invalidates most of the rest of the discussion in Unwin's book. If you are not convinced, consider the following:

Proposition K= Kukuriku exists

Do you think proposition K is true or false?

Clearly, there are more than two answers to this question.

On page 48, the author discusses the case of tossing a coin with the assumption that there are two possible outcomes: head (H), and tail (T). In fact, one must take into account other possible "events" or outcomes. For instance, that the coin will fall on its side, or that the coin will break into pieces, and it would be impossible to determine if it is an "H," or a "T." We normally assume that in this case there are only two outcomes with significant probabilities.

Let us go back to the proposition G. It is not clear what $P(G)$ is, what $P(G|E)$ is, and what $P(E|G)$ is. This is sufficient to render the whole application of the Bayes' Theorem unwarranted, if not meaningless.

Next, on page 58 the author makes an assumption that a good starting guess, based on complete ignorance would be 50-50 argument.

Thus, his assumption is that:

$$P(G) = P(G^*) = 50\%$$

He comments further that "*this is the perfect, unbiased expression of agnosticism.*"

The fact is that this is a *perfect*, *biased*, and *deceptive* assumption. If you are not convinced, consider the proposition

$$G = \text{Kukuriku exists}$$

$$G^* = \text{Kukuriku does not exist}$$

Would you accept the assumption, that the probability of $P(G) = P(G^*) = 50\%$, is the "perfect, unbiased expression of agnosticism?" Clearly, this is a meaningless assumption if you do not know who or what "Kukuriku" is.

Obviously, the author draws this assumption from the experiment of tossing a coin. If we assume that there are only two possibilities: H and T, and that we have no reason to suspect that the coin is unfair or unsymmetrical with respect to H and T, then it is reasonable to expect that the initial probabilities (i.e. before having any new information) are: $P(H) = P(T) = 50\%$.

The reason for this particular choice is that we can support it by either using the symmetry argument with respect to H and T (see classical definition in Note 1), or actually tossing the coin many times, and recording the relative frequencies of the occurrences of H and T. Neither of these possibilities are at our disposal in the case of proposition G. First, we cannot apply the symmetry argument for G and G^*. Second, we cannot make experiments and count the frequencies of appearance of G and G^*. And finally, in the case of a coin we all agree on what H and T are, and we also know what a coin is. In the case of G, not everyone knows what God is (and by the way, if one knows what God is, he/she does not need any probability of G).

7. Chapter 5:   The Bayes' Craze

The chapter opens with two questions

"What would be some practical applications of Bayesian inference?

Is it technically legitimate to represent degrees of human belief in mathematical form?"

The author gives several applications of Bayes' Theorem to give the reader a "flavor for the breath of Bayesian applications." None of these are discussed in any detail [for a few legitimate applications of Bayes' Theorem, see Ben-Naim (2015b)].

It seems to me that the author provides this list of examples to lend credence to the case in which he applies the theorem. The author realizes that this particular case does not fall in the same range of topics that he mentioned. On page 63, he even makes a correct statement, that the very idea that one can mathematically represent the workings of the human mind, has a certain arrogance to it.

Indeed, the very idea of applying the Bayes' Theorem to the "workings of the human mind" has a considerable degree of arrogance. However, using the Theorem for the proposition G is far more than mere "arrogance." It is foolish!

On pages 69-74 the author discusses a vignette in which the author and Frankie (the Frequentist) are discussing the questions posed in the beginning of the chapter. One statement shoved in Frankie's mouth is important:

"There is no such thing as the probability that God exists, so how can I estimate it? Probabilities come from repeated trials of events, like the coin tossing you've been deriving boundless fun from. Proposition G has nothing to do with repeated trials. You know, sometimes the appropriate conclusion is that there is no meaningful question to answer, and so trying to answer it is a vacuous exercise."

This is true. There is no way to assign any probability to proposition G. Yet, the author promises to Frankie that he will produce a probability that Proposition G is true. Adding that:

"You may not like some of the assumptions or conclusions I make along the way."

It is not a question of *liking*, or *not liking* the assumptions, and the conclusions. Both are absolutely meaningless, and therefore one should *dislike* and *distrust* the entire book.

This chapter is concluded by begging for answers to the questions posed in the beginning. It is merely a repetition stated earlier that Bayes' inference has proven to be a powerful analytical tool in a range of technical descriptions. Again, this is true. Unfortunately, the author fails to mention that proposition G does not belong to any technical discipline!

8. Chapter 6: Down to business

Notwithstanding the questionability of using the Bayes' theorem for proposition G, the author gets to do his business; in this chapter he discusses the possibilities of evidence to support or refute the proposition G. Before describing the author's own "evidentiary areas," let me ask the reader to consider the following exercise:

"Proposition G. The existence of a dog having fifteen legs and five eyes."

If you are agnostic, a fair, "perfect, unbiased assumption of the prior probability would be:

$$\text{Probability of } G = 50\%$$

Now, consider the "new evidence" that I tell you as a *fact*, and you accept is as truthful

E

= I know of a dog who saved the life of a little girl who fell into the poo

How would you revise or modify your prior probability of G, given the evidence E. Only a qualitative answer is required.

A more challenging exercise: How would you change your estimates of both P(G) and P(G|E), if I replaced "dog" in the previous exercise with octopus?

Now that you have done the exercise let us go back to the main problem of this book. In order to use the Bayes theorem, we need to define some evidence.

Before listing his "evidentiary areas" the author reminds us of the proposition G. He also clarifies that he means "the God of the major monotheistic faiths."

I doubt that the "God" of the major monotheistic faiths is the same God, and besides, what about the other Gods of other faiths?

Then, the author describes some attributes of the God he had just defined:

"The person-God is compassionate, fair, and merciful. He is concerned with each and every human life and sustains us individually. We can communicate with him personally, through prayer or perhaps through deed..."

The author forgot some attributes that are explicitly discussed in the Bible, e.g. that God is revengeful, and that He punishes those who worship other gods, and He will also eliminate and destroy all those who do not believe Him. If you read the Bible carefully, you will also find that God is sometimes childish, sometimes arrogant, and sometimes cruel. Whatever your god is, it is hard to accept that all the attributes, given by the author are accepted by all believers. Besides, all those attributes have nothing, absolutely nothing to do with evidence, or with the probability of G. These are attributes assigned, quite arbitrarily by some people to some gods. Let us go through the list of evidentiary areas that the author uses as "evidence" to reassess his probability of G.

Evidentiary area: 1. The Recognition of Goodness

The author starts with some banal description of what is good, and what is evil:

"We humans seem to have a sense of what is good and what is evil. We certainly encounter areas of ambivalent gray in our lives, but even the

recognition of grayness requires an understanding that is a mixture of black and white components, a mixture of good and evil. "

I doubt that every human being has the same sense of "good" and "evil," or recognize the gray "mixture of good and evil."

Once again, he explains the insipid idea that the indifferent-world of physical systems – meaning non-living systems – have no place or language for goodness.

This is why "goodness" is not encountered in a course on electromagnetism.

Great! The author forgets that "goodness" is also not generally (actually never) a learning unit on a course on probability, or any other course in science!

Then as "evidence," that goodness must have originated from God, he quotes Genesis 1:27:

"So God created man in his own image, and so innate in us is a recognition of the good."

Of course, even without calculating anything (as he will do in the next chapter), it is clear that if you believe in the Bible, and if you believe that the origin of goodness is in God, then you must also conclude that the fact that goodness exists supports the Proposition G. The author forgets however, that if you believe in the Bible, and you believe that goodness originated from God – this already implies that you believe that Proposition G is true, and therefore you *do not* need any proof of Proposition G, neither from Bayes' theorem, nor from any other mathematical theorem. This kind of futile arguments will go on and on throughout the rest of the book.

At some point, the author asks a rhetorical question:

"Did what you just say have any meaning, and if so, what is it?"

The author does not answer this question. Instead, he quotes some philosophers and psychologists who wrote about goodness being relative, and that some minorities such as Communists, Jews, and homosexuals were purged, expelled, or killed by "cohesive moral communities."

He wraps up the section by claiming that "goodness" is an important area of evidence.

In my view the "recognition of goodness" has nothing to do with the truthfulness of Proposition G. It is also frivolous and inappropriate to use this "evidentiary area" in Bayes' Theorem, as he will do in the next chapter.

2. Evidentiary area: The existence of Evil

There is no point in going through the arguments provided here for the effect of the "Existence of Evil" on the probability of Proposition G. As I explained before, if you believe in God, and if you believe that the origin of goodness is in God, then you do not need any proof of proposition G. If you are aware of the existence of evil it would not affect the extent of your belief in God. As we shall see later in the next chapter the author uses the "Existence of Evil," in Bayes' Theorem to *reduce* the probability of G. This type of argument reminds me of two stories in which a similar argument is involved.

During one of my sabbatical years in the US, I lived next door to the rabbi of that community. One day the rabbi's daughter nearly drowned and almost died. After an intensive treatment – and perhaps an additional "miracle" – the daughter recovered. When she came home, her father, the rabbi said they have to go to the synagogue and *thank* God for saving his daughter. Upon hearing this, his wife instinctively asked; what if our daughter died, would we also thank God or perhaps would have cursed him instead?

Obviously, those who believe in God's goodness will always *thank* God for whatever happens. If something bad or evil happens it would not affect their belief – they will always relegate the cause of the evil to someone or something else, or true to their faith, the unfortunate event as God's will.

The second story is quite familiar, having heard it many times when a religious person argues with a non-religious one. The non-religious person asks, "How can you believe in God after what the Jews went through in the Holocaust?" To this, the religious person answers, "You have to thank God that the Holocaust happened, because as a result of that the State of Israel was created."

Another more recent "justification" of the Holocaust is that God punished those Jews because of their sins. Ironically, even unborn babies in their mother's wombs, and little children were punished by God because of their sins. What a distorted view of God's goodness!

Thus, whatever happens, be it good or evil, for as long as one's faith is steadfast and unbreakable, nothing will hardly have any dent on one's belief. There are exceptions, of course. I heard more than once such a declaration:

"I was a believer all my life until I saw the horrors of the Holocaust. That was the straw that broke the camel's back."

3. Evidentiary area: Miracles

The author starts with the assertion that "Miraculous events" play an important role in most versions of the Judeo-Christian faiths. And that the word "miracle" is used in various senses.

After a lengthy discussion and explanation of what a miracle means, the author distinguishes between two classes of miracles:

1. Phenomena that could be construed as being brought about by natural forces but were in fact initiated or facilitated in some way by God. An example might be remission from a serious disease due to an act of God.
2. Phenomena that violate natural laws and indicate a blatant suspension of the physical workings of the universe. Raising the dead would be an example.

The first is referred to as "*Intra-Natural Miracle*," and the second as "*Extra Natural Miracle*." These two will be treated separately in the next chapter to reassess the probability of proposition G. In my view both of this "Natural Miracle" are no more than juvenile interpretation of the word "miracle." Remission from a disease can or cannot be "initiated or facilitated" by some superpower one chooses to call God. By the same token, raising the dead can or cannot be "initiated or facilitated" by some superpower. We do not know any of these!

4. Evidentiary area: Religious experiences

Everyone has experienced a sense of awe and a sense of wonderment by just looking at the vastness of the night sky, or at a multihued sunset. The interpretations of such experiences, in terms of "feeling one's standing in relationship to the divine," depends on whether one is a believer or not. I have had that experience so many times but it never occurred to me to consider those as "evidentiary area."

This experience will also be used as evidence in Bayes' Theorem in the next chapter. In my opinion, all these "evidentiary areas," are very vague, ambiguous, and their interpretations depend on whether one is a believer or not. None can be assigned any probability, and none can be used in the proper Bayes' Theorem to either enhance or diminish the probability of proposition G, a prior probability that does not exist in the first place.

Besides, the choices of these particular four evidentiary areas (two of which are further split into two – see next chapter – so they make six) are totally arbitrary. I could add any other experience or phenomena to the list above: The Holocaust, the existence of the Bible, the fact (?) that Moses met God, the fact that love exists, the existence of humans – and you can say anything you want to plug into the Bayes' Theorem to obtain any number you want to get. I shall give you an example of my own calculation after discussing Chapter 7, and also suggest to the reader to construct his/her calculation based on Bayes' Theorem.

This chapter ends with the following meaningless sentence:

"In the next chapter, the rubber hits the road and the numbers fly. Put fresh AAAs into the TI, sharpen the 2HB, and blow clean the Snoopy eraser: It's math time!"

No! This is not math, but pseudo-math, or better yet deceptive math.

9. Chapter 7: The Numbers

The purpose of this chapter is to calculate how each evidentiary area affect the truth probability of Proposition G.

The six "evidentiary areas" that the author will consider are:

1. the recognition of goodness
2. the existence of moral evil
3. the existence of natural evil
4. intra-natural miracles
5. extra-natural miracles
6. religious experiences

These are denoted by E1, E2,…, E6. The plan is to use these evidentiary areas *serially*, i.e. first, consider E1, then E2, and so on. I should emphasize right now that this list of "evidentiary areas" is *totally arbitrary*. You can add whatever comes to your mind to this list. Suggestions: Existence of matter, hating experiences, intra-natural love making and extra-natural love making. The reader is urged to suggest his/her own "evidentiary areas" and use them later within the Bayes' theorem. Of course, none of these "evidentiary areas" are evidence of anything, and would not affect the "truth probability of proposition G."

In the beginning he starts with the prior probability of 50% which was discussed in Chapter 4, and which is a totally meaningless assumption. You can start with any number you want; it will equally be meaningless! Another unjustified assumption that the author makes on page 85 is that evidence items E1 through E6 to be mutually independent in the sense that ..."

This is as obscure as any of the other assumptions made throughout this book. It is obscure in the sense that independence in probability means that occurrence of one event does not affect the probability of a second event. For instance, if we throw two dice which are very far from each other, then the occurrence of "4" on one die has no effect on the probability of "6" on the second die. These two events are said to be independent. We write this as either

$$P(\text{"4" on one die } and \text{ "6" on the second die})$$

$$= P(\text{"4" on one die}) \times P(\text{"6" on the second die})$$

On the other hand, the occurrence of "4" on one die makes the probability of "6" on the *same* die zero. This is a very strong dependence. We can rewrite these two cases in terms of *conditional* probabilities:

For two (very far) dice

$$P(\text{"6" on one die}|\text{"4" on the other die}) =$$

$$P(\text{"6" on one die})$$

For the same die $\quad\quad\quad P(\text{"6"}|\text{"4," both on the same die}) =$

0

Now, let us check what it means by "existence" of goodness" is dependent or independent of "existence of evil." In fact, I have no idea what probabilities one can assign to these propositions. Either we know (or believe) that goodness exists or not. The same is true for any other pair of evidentiary areas. For instance, I know that a miracle has occurred, how does that knowledge affect the probability of the "existence of goodness?" The reader is urged to pause at this point and think whether it is worth to continue reading all the details of the silly-to-meaningless derivations that follows. I have to continue to read in order to write this review. By the way, I had the same experience while reading all the books I have reviewed in these series of books.

Besides the dubious meaning of the mutual independence between E1 through E6, the author makes a technical mistake in writing the equation:

$$P[(E1 \text{ and } E2)|G] = P(E1|G) \times P(E2|G)$$

This means that *given* G, then E1 and E2 are independent. Thus, what he refers to as mutual independence between E1 through E6 is actually mutual independence between the *conditional* events (E1|G) through (E6|G) . This means that if you think (know or believe) that G is true, then the occurrence of "goodness" is independent of the occurrence of evil (or miracle or whatever).

With this nonsensical assumption of independence, the author writes the general formula of computing P_{after} given P_{before}

$$P_{after} = \frac{P_{before} \times P(E|G)}{[P_{before} \times P(E|G)] + [100\% - P_{before} \times P(E|G^*)]}$$

This is the Bayes' Theorem. The author correctly points out that by merely looking at this equation, you can see that once we have the starting probability P_{before}, the two quantities we need in order to calculate are P_{after}, are $P(E|G)$ and $P(E|G^*)$.

This is true. Unfortunately, none of the quantities are known. Specifically, we do not know the ("strictly mathematical") probabilities P_{before}, $P(E|G)$ and $P(E|G^*)$. Therefore, this equation is *useless* for the calculations he is going to do in the entire chapter.

Thus, whatever arbitrary meaningless number you plug into the right-hand side of the equation, you will get another meaningless number on the left-hand side; in short, Garbage in, Garbage out (GIGO)[3], Figure 2

Figure 2. GIGO: Garbage in, garbage out

The author is well aware under which conditions one can use this formula. On page 96, he correctly writes that for proper applications of Bayesian inference, one needs hard statistical data.

Then, he admits that Proposition G does not have this sort of statistical data.

Therefore, in effect the author implicitly admits that all the numbers he will use in the following calculations are arbitrary numbers hence, the results obtained are *senseless*!

Notwithstanding this admittance, the author (nor anyone else) has any "strictly mathematical" probabilities to use in Bayes' Theorem. The author continues to make another extraordinary misleading trick to convince the reader that his calculations are meaningful. First, he defines a quantity

$$D = \frac{P(E|G)}{P(E|G^*)}$$

which he refers to as the "Divine Indicator," explaining that the greater its numerical value, the more the evidence enhances the probability of the existence of God. This explanation of D is incorrect even when the quantities $P(E|G)$ and $P(E|G^*)$ are proper (*strictly mathematical*) probabilities. It becomes meaningless for the quantities $P(E|G)$ and $P(E|G^*)$ as "defined" in this book. Moreover, the very choice of the term "Divine Indicator" might mislead the reader that there might be something "divine" in this quantity. I would suggest to use the word "adivinar" which means "guessing" for these numbers.

Finally, to give further (false) credibility to the quantity D, the author provides a Table of D Values and comments that this table is "similar to the "Richter scale" for earthquakes and "Fujita scale" tornadoes.

What Richter did for earthquakes, and Fujita did for tornados have nothing to do with the laughable notion of "God related" evidence. This is pure misleading baloney!

Then we get the corresponding Table D on page 100. It is so absurd that I urge the reader to look at it and enjoy it.

Instead of selecting any arbitrary numbers for D, he chooses to select these particular values and refer the reader to a lengthy appendix (see page 233) on the "Mathematical Theologist's Spreadsheet."

The seven-page long appendix is totally deceiving, concluding with a reassurance to the reader that now, after reading the appendix he/she is a "mathematical theologist."

So, all these mumbo jumbo about the "Divine number" has transformed you into an overnight sensation, elevating you into a position better than Aristotle, Thomas, and Kant, perhaps better equipped to do the following GIGO calculations. By the way, there is no such thing as "mathematical theology." It is plain "theology" cloaked in a mathematics t-shirt!

On page 101 the author explains why he uses these particular numbers for the Divine factor D. He says that 10 will be the limiting value of the Divine Indicator; why not 100, or 1000?

This reminds us of his prudish assessment of the accuracy of his calculations (note four but two significant digits...). The fact is that any arbitrary number you choose for D will have the same effect in producing the same garbage!

Let us now go through the sequence of re-evaluating the probability of G for each evidentiary area.

E.1   Recognition of goodness

This section starts with a meaningful, but unanswerable question, namely:
In a godless universe, would goodness have meaning?

Everyone is supposed to know what *goodness* means, whether one believes in God or not. For those who are believers, a "godless universe" is meaningless, for unbelievers, God itself is meaningless.

Following this question, the author fills a few pages with philosophical scribbles about the Golden Rule ("Do unto others as you would have them do unto you," page 103), or quoting the Dalai Lama ("All humans desire to be happy and avoid suffering"). Included in his litany of quotations and personalities is Mother Teresa, as well as many others in order to justify his choice of the Divine Indicator for this evidentiary area to be $D = 10$.

Not only is this number arbitrary – if not meaningless, but after his lengthy philosophical discussions the author refers the reader to Table D, claiming that this is precisely the expression of relative likelihood, corresponding to a Divine Indicator value of D= 10.

Who can argue about the *precise* number taken from the "Divine Table D?" If you still believe in the Divine Table, you should continue reading the book. Otherwise, there is no point in continuing the reading of this book.

Now that we have a "reasonable numerical value of D, together with the "reasonable" assumption about the prior probability $P_{before} = 50\%$, we can plug these numbers in Bayes' Theorem and get:

$$P_{after} = \frac{D \times P_{before}}{D \times P_{before} + 100\% - P_{before}}$$

The author triumphantly pronounces that after this calculation the probability of "Proposition G" increased from 50% to 91%:

"A good start for God!"

For the lay reader I should add a comment here that will apply to all subsequent calculations: Neither the value of P_{before}, nor the value of D has any meaning. Therefore, the result 91% is not "the truth probability of proposition G," but the first arbitrary meaningless number. I will add one comment to the "believers." If I were God, I would have been insulted by assigning any probability on my existence. Therefore, the result obtained is not a "good start for God."

E2:   The existence of moral evil

After a lengthy philosophical-for-a-penny discussion, the author decides that the Divine Indicator – taken from the "Divine Table D" is D = 0.5. He plugs this number into Bayes' Theorem using the P_{before} = 91% to obtain; "the truth probability of proposition G" is

$$P_{after} = 83\%$$

Commenting cleverly following this result "God has sustained a minor blow."

Whatever the meaning of the result he obtains, it has no effect on God, or on the Probability of God, or on anything. This is pure nonsense. Again, if I were God I would not accept the 91% in the first step, nor would I consider it as a "minor blow" in the second step! If I were a believer in God, I would have been insulted by all the calculations made in this book, including the final result obtained by Unwin!

E3:   The existence of natural evil

Starting with a startling list of natural hazards: earthquakes, tornadoes, cancers, etc. the author yet again, asks an unanswerable question:

"Why would the omniscient, omnipotent, good God allow these natural hazards and their frequently horrific consequences?"

Although no answer is provided the author reaches an important, though silly conclusion again, after consulting the "Divine Table D, providing the value of 0.1, and using this value of D, the "truth probability" drops from 83% to 33%. What an unpleasant blow to God! In my view it is a blow to the author, rather to God or to anyone else. Besides, how does the author know that "this evidence disfavors God?"

E4:   Intra-natural miracles

Examples: "Remission of an illness, safe landing of planes, etc." and three more pages of scribbles, and consulting the "Divine Table D" leads to the (meaningless) choice of $= 2$, plug it into the Bayes formula, and one gets a new "truth"

$$P_{after} = 50\%$$

What a pitiful situation, after nearly 20 pages or "deep arguments" and "sophisticated" mathematical calculations we come to full circle, the original (meaningless) P_{before} we started with! Poor God, so many blows in one book.

E5:   Extra-natural miracles

This section starts with an honest statement that he, the author never witnessed an extra-natural miracle.

Then, yet another unanswerable question:

"If such miracles do not occur, what would account for the reports?"

His answer is:

Deception and self-deception are the obvious explanation.

This is true; not only about this particular evidence (that no one has witnessed) but also to all other evidence, as well as to the whole book on the "Probability of God."

But the author is unstoppable as he continues to consult the "Divine Table D," and gets a number $D = 1$, which as you might guess has no effect of the "truth probability," i.e. $P_{before} = 50\%$, transformed into $P_{after} = 50\%$ which means that the evidence is "God-neutral."

What an enlightening revelation! This is of course true for all other evidence.

Let us turn to his final evidentiary area.

EG: Religious experience

As a young boy, I have experienced what the author refers to as religious experience, but I do not know how to describe that feeling, nor could I define or quantify it. All I can say is that after my divorce from God at age 14, this religious experience went out the door, and never came back.

Many years later I have heard people talking about their religious experiences, and I vaguely understood what they meant until I talked with a mathematician who happens to be religious, but strongly believes in the theory of evolution. I asked him whether he does not see any conflict between believing in God, the Bible's story about creation, and Darwin's theory of evolution. His answer; my *religion is an experience not a theory* [in Hebrew it sounds slightly different:

‏["בשבילי הדת היא חוויה, לא תורה"].

What he meant is that religion is not a theory of anything, but rather a feeling or experience, and as such, it does not conflict with any theory.

As always, before executing the actual calculation the author presents a very erudite discussion on the question of whether religious experience could, or could not exist in a universe, with or without God. After two pages of scribbles on this question, he proceeds to consult with the oracle in Table D and comes up with the moderate Divine Indicator of D = 2, plug this number in Bayes' equation, and get P_{after} = 67%. Then, he concludes that the final truth probability of Proposition G to be 67 percent.

The number 67% is explained as; the odds are 2 to 1 in favor of God. This number is referred to as "magical number" lending to it not only a "true" value but a "mystic" attribute as well.

Before we go to the next chapter we should pause here and summarize what we have achieved.

We started with the initial probability P_{before} = 0.5 (or 50%). Let us call this quantity GI1 (for "Garbage In" number One). This GI1 is used as an input in BF (Bayes' Formula), using a Divine factor D1 to obtain GO1 (Garbage Out number 1). Schematically, we write this first operation as

$$GI1 \xrightarrow{D1} GO1$$

Next, we take GO1 and rename it as GI2. This means we take the garbage *output* from the first step and use it as *input* garbage for the second step which is schematically described as

$$GO1 = GI2 \xrightarrow{D2} GO2$$

This process is continued four more times:

$$GO2 = GI3 \xrightarrow{D3} GO3$$

$$GO3 = GI4 \xrightarrow{D4} GO4$$

$$GO4 = GI5 \xrightarrow{D5} GO5$$

$$GO5 = GI6 \xrightarrow{D6} GO6 = 67\%$$

Clearly, in the six steps of GI → GO, we transferred the initial number of GI1 = 50% into the final number GO6 = 67%. It should be emphasized that all these garbage numbers are valid when we know what the meaning of GI1 is, as well as the meaning of all the "Divine" quantities D1, D2, D3, D4, D5, D6. If the reader is not well-versed, as I readily admit to be one, to know what GI1 is, and what all the D-values mean, then all the garbage numbers turn into senseless arbitrary garbage numbers.

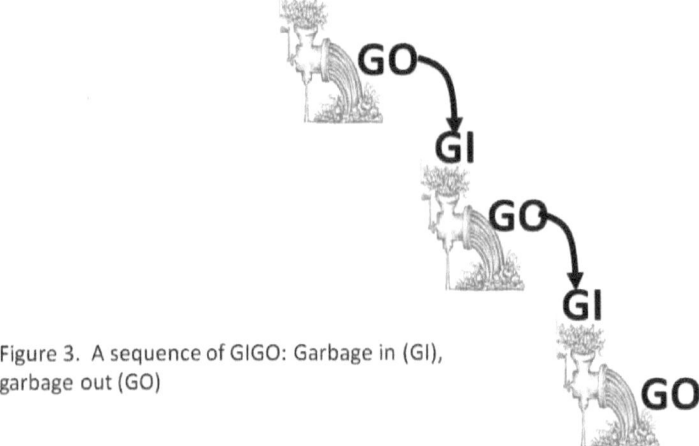

Figure 3. A sequence of GIGO: Garbage in (GI), garbage out (GO)

So far, I have described in simple and comprehensible terms what Unwin did in this chapter. I want to add one criticism of Unwin's procedure. Why stop after six steps? The author is honest enough to suggest to the reader who might have "new evidentiary areas" an appendix where a "computer inclined reader" can find an "electronic spread-sheet to facilitate the calculation."

The problem with this "Theologist's spreadsheet," as described in the appendix on page 233 is that if you invent new evidentiary areas you will only get new garbage. Besides, you will never know when to stop the process. For any GIn (Garbage In at step n) there is another $GI(n + 1)$. Hence, by mathematical induction, you could spend your entire life transforming GI into GO.

10. Chapter 8: Risk Apocrypha

In the opening sentence of this chapter the author admits that the last estimate of the probability that "God exists" (67%), has a subjective element, because it reflects his subjective assessment of the evidence.

Subjective element? I thought the author was planning to do "strictly mathematical probabilities." Did he forget his promise in Chapter 1? I would like to refresh your memory about his promise to use the term "probability "in its strict mathematical sense"

The chapter starts with what is known as Pascal's wager which is "an interesting idea about the role of probabilities in religious decision making," then discuss some obscure topics such as "afterlife," "God-pleasing life," and "God-displeasing life," and about the assumption made by Pascal that the gain of living a God-pleasing life, if God exists, is infinite.

After all these gibberish, the author reaches infinity, symbol: ∞. He mentioned some interesting arithmetical operations, such as:

$$\infty + \infty = \infty$$

$$\infty \times \infty = \infty$$

$$\infty \times 0 = ?$$

Then, he concludes the chapter with questioning the 67% result he got, and whether he could not have produced similar results simply "off the top of my head had."

If all he could hazard a guess was a mere 67% without using the Bayes Theorem, why did he write this book in the first place? In my opinion, this sentence is an admittance that all he had done so far was purely guessing, "adivinar," in Spanish, and the whole story about the Bayesian analysis was pure nonsense at best.

The author promises to discuss the answer to this question in the next chapter. Before we turn to the next chapter I urge the reader to do the following exercises with *arbitrary numbers*, symbolized by α.

Calculate the result of the following operations with α.

$$\alpha + \alpha =?$$

$$\alpha - \alpha =?$$

$$\alpha \times \alpha =?$$

$$\alpha/\alpha = ?$$

Remember that α stands for an *arbitrary number*. Compare your results with the note 4.

11. Chapter 9:   Probable Thoughts

The question posed for this chapter is:

"What are the perspectives of Chad and the Axe, the mall intellectuals?"

After eight pages of meaningless scribblings, he concludes:

"Think carefully before you have your body pierced."

Since reading this chapter left my mouth agape, and scratching my head not knowing what this is all about, I will refrain from commenting on it.

12. Chapter 10:   Faith Math

This chapter poses four questions which are equally meaningless. Here is a representative question: Can an atheist experience faith-based belief?

Since I am clueless as to what experiencing "faith-based-belief" is, and since this question is equally meaningless to the other three questions posed in this chapter, I must admit that the meaning of this chapter is lost on me, and therefore I am not qualified to comment on it.

Reading through these unintelligible 25 pages, we are treated to a most unexpected surprising conclusion, that his own assessment of the probability of "Proposition G" is 95%.

You can forget about the 67% he got after so many calculations based on Bayes' Formula. The new, true (strictly mathematical?) value of the probability of proposition G is 95%. Great! Thus, finally he admits that his own degree of belief in proposition G is 95%. If I were God, or any believer I would feel utterly insulted. The mathematics led him to 67%, and his own (objective, subjective, mathematical???) is 95%. As God I would not accept any number below 100%.

13. Chapter 11:   An existing question

Remember the statement: "I did not have sexual relations with that woman," from Chapter 2? Only now, after pages upon pages of calculating the probability of proposition G – which is "God exists," the author remembers that one needs to define the word "exists" (presuming that we know what the words "proposition" and "God" mean).

Thus, the question posed for this chapter concerns the meaning of the word "exists" in his analysis of the probability of the "proposition that God exists"

The author admits from the onset that after that long "journey" he feels a sense of incompleteness, and perhaps also a "feel of guilt."

Indeed, he should have a "feel of guilt" after forcing the innocent reader to go through all this Garbage.

The author goes on to discuss the difference between the existence of a mathematical object and an abstract concept, between material things and thoughts.   All these philosophical questions and pseudo calculations of a "strictly mathematical probabilities" have no place in a book on "Probability of God".

Then the author asks the reader to choose between two absurd options:

The first option: To live in a universe in which "God exists," but a universe in which only a few take notice of Him.

The second option: To live in a "godless universe," but a universe in which everyone erroneously believes that He exists.

I suggest to the reader to "invent" other interesting options. This will surely be an entertaining exercise.

Then he ponders on an even more absurd question; which of the two options mentioned above, God himself would choose for us…

Doesn't the author know that God has no choice? He must use Bayes' theorem to find out what the "probability" is of each of the options!

And finally, the trivial suggestion to the reader that whatever his or her intuition tells them about the meaning of "existence" one should stick with it.

This renders the whole chapter, as well as the entire book superfluous.

14. Chapter 12: Faith AfterMath

The next and the last chapter is on: "Faith After Math." It contains more irrelevant "philosophical" discussions on *faith*, *probability*, and faith as a possible source of evil.

It starts with the question:

"What are the collateral benefits of quantitative analysis?"

If by "quantitative analysis" the author means the analysis made in this book, then the answer if straightforward; no benefits at all. Then comes another question:

"Can faith be a source of evil?"

The painful answer to this question is: Yes, of course. And we witness this kind of evil every day.

There is one more moving story which though clearly irrelevant to the "Probability of God," mitigates the extremely boring character of this chapter. The story is taken from Karen Armstrong's book on the History of God (1994). It is about events that occurred in the Auschwitz concentration camp in which a group of Jewish prisoners put God on trial. God was charged with cruelty and betrayal. Like in the Biblical story of Job, they found no consolation in the usual answers to the problem of evil and suffering in the midst of this current obscenity. They could find no excuse for God, no extenuating circumstances, so they found him guilty and worthy of death! Then the Rabbi pronounced the verdict. The trial was over.

In my opinion this last short story should have *replaced* the entire book.

The book concludes with the statement:

"I'm thankful for this balance. So these are the humbly offered thoughts of a 95 percenter. But one day I think my number will be up."

Up? With what new "evidentiary areas?"

Exercise: Use the Bayes' theorem to prove the probability that Kukuriku is 95%, or any other probability you want. See calculation in Note 5.

Some concluding remarks

This book does not add a feather to the author's cap. It is an unflattering book which is a disgrace to science. In the book's back cover, we find:

"This groundbreaking book reveals how a mathematical equation developed more than two hundred years ago can be used to prove the probability that God exists."

This is not a groundbreaking book, but rather a "law-breaking" one which reveals how a scientist with a PhD in theoretical physics can *abuse* and *twist* an equation developed more than two hundred years ago in order to peddle the idea that he has "proved" the "probability that God exists."

It therefore came as a shock to me to read Michael Shermer's review of the book which is found in the back cover:

"One of the most innovative works [in science and religion movement] is The Probability of God…An entertaining exercise in thinking."

I had great respect for Michael Shermer, the "skeptic." This comment on the cover of the book made me *skeptic* about Shermer's skepticism!

Part VI: Review of:
Proof of Heaven
A Neurosurgeon's Journey into the Afterlife
By:
Eben Alexander, M.D.

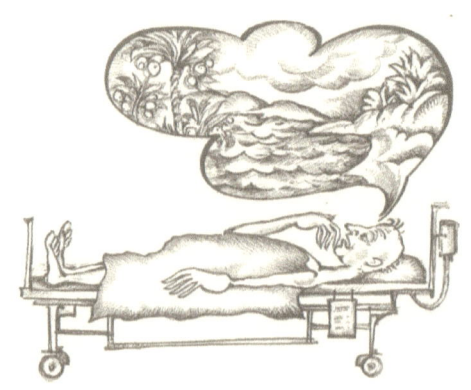

1. Introduction

This book is written by a neurosurgeon who writes on a nonscientific topic. The author misuses and abuses the concept of "Proof," thereby misleading the reader into thinking that the content of the book is "scientific" or "trustworthy."

2. The title: Proof of Heaven

The book's title is not only deceptive, it is also a misuse of the term "proof" as nowhere in the book can be found a proof of anything, and there certainly is no proof of heaven. Moreover, the term "proof" is abused as it was used in connection with a concept, which in principle, is unprovable. As I have commented on the title of the book "Probability of God," there is no such thing as "proof of heaven" as much as there is no "proof of Hell," or "proof of a chair." What the author wants to say is that Heaven is real or that his experience "proves" that Heaven exists. Unfortunately, the whole story told in the book *proves* only one thing; that the author does not know what the word "proof" means. See Note 6.

3. The subtitle: A Neurosurgeon's Journey into the Afterlife

The subtitle reinforces my view about the inappropriateness, and misleading nature of the title. First, the author with a science background uses the "proof" in a book where no proof is presented. Second, the book does not offer a "journey into the afterlife." As far as I understand, the term "afterlife" means after-life. Equivalently, "afterlife" is supposed to be after being dead. The author was never pronounced dead, therefore his narrative about the "journey," was not a journey, and was remotely close to the concept of afterlife.

On the book's cover we find a comment by Raymond A. Moody:

"Dr. Eben Alexander's near-death experience is the most astounding I have heard in more than four decades of studying this phenomenon. [He] is living proof of an afterlife."

This is not true. The book is about "a near-death experience," not afterlife or after death. I heard and read several NDEs, none of them astounding, including this one!

On the back cover, repeated thrice was the reference to the author as a scientist, and as a doctor. In addition, we read:

"Alexander's recovery is a medical miracle. But the real miracle of his story lies elsewhere. While his body lay in coma, Alexander journeyed beyond this world and encountered an angelic being who guided him into the deepest realms of super-physical existence. There he met, and spoke with, the Divine source of the universe itself."

Whether an NDE may be viewed as a medical miracle depends on one's belief in miracles. The fact that this "miracle" happened to a "highly trained neurosurgeon" does not lend any credibility to either the story he tells us, or to the interpretation of the story. At most, one can say that this particular NDE is banal as any NDE, childishly interpreted.

The statement that "Alexander's story is not a fantasy" is also misleading. Finally, we find in the back cover:

"This story would be remarkable no matter who it happened to. That it happened to Dr. Alexander makes it revolutionary. No scientist or person of faith will be able to ignore it. Reading it will change your life."

Such stories happen to almost anyone who had ever *dreamt*. There is nothing remarkable in that (I am referring to the "story," not the actual recovery of Dr. Alexander). The fact that it occurred to Dr. Alexander is not revolutionary, contrary to what is written in the back cover – in any sense one might interpret the term "revolutionary." Why it should be considered revolutionary is baffling. For all those who have experienced NDE, the profession or their simple status in life does not really matter; it does not make the experience more special just because of one's title or profession. He is a neurosurgeon who also happens to have a knack for creative writing, but I will not diminish the NDEs of the simple folks who probably saw the same beautiful things but do not have the talent as the author has, or simply did not see any need to write a book about their experience.

I also believe that any person (scientist or not), *should* not take such story seriously, as much as I have ignored so many times such stories in my dreams.

Finally, reading this book had changed *nothing* in my life. If at all, the only effect on "my life" was the decision to write this book-review.

4. The Prologue

The prologue opens with one of Albert Einstein's famous quotations:

"A man should look for what is, and not for what he thinks it should be."

I agree. Einstein's message applies to Alexander's book. Anyone who reads and understands Einstein's statement should not write such a book.

The prologue essentially contains two parts. One tells some biographical events about dreams and parachuting, and one statement that I fully agree with, that the brain is an "extraordinary device," more extraordinary than we can imagine.

The second part seemingly aims at convincing the reader that as a neurosurgeon (as we are repeatedly reminded: "I am a neurosurgeon") the author is qualified to analyze how his own brain works, including his extraordinary experience. He then tells us what occurred on November 10, 2008. He was struck by a rare illness and was thrown into a coma for seven days. Then, there a technical description of his state. His entire neocortex shut down. Inoperative. In essence, absent.

I understand what he meant by saying that his brain "shut down" while he was in a coma. I can also understand what he meant by "Inoperative." What I do not understand however, was his conclusion that his brain was "absent."

Then, there is a lengthy story of his "conversion." Initially, unbelieving in other people's stories who had "strange experiences," such as stories of traveling to mysterious places, of talking to dead relatives, and even meeting God himself.

In his opinion, those stories were "pure fantasy," until he himself had a near-death experience (NDE). This is a "standard" trick used by many religious leaders with the usual line saying that when they were young and unwise, they did not believe. But now that they are old, they are wise, experienced, and are "converts" to believing. My story is different; when I was young, I was a believer, once I opened my eyes and mind, I converted. I would never use this story to convince anyone to believe or not believe.

In his case, it was not merely "pure fantasy" but a reality. There is also a paragraph which tells us that Dr. Alexander, a neurosurgeon, with decades of research, was in a better-than-average position to assess and judge the reality as well as the implications of what happened to him.

I am not convinced that he, or any other scientist for that matter, who have experienced NDE is in a "better-than-average" position to judge the "reality" of what happened.

After his NDE, he was convinced that "death of the body and the brain are not the end of consciousness," and "human experience continues beyond the grave."

Since he was not really dead, and was not "beyond the grave," it is unconceivable how Dr. Alexander could have reached those conclusions.

These statements clearly discredit the author's interpretation of his NDE. He even goes further by concluding that after death (which he had not experienced), the conscious:

"continues under the gaze of a God who loves and cares about each one of us."

At this point, it is clear that what Dr. Alexander is going to tell us should not be read and interpreted as something written by a scientist, but rather by a religious preacher who wholeheartedly believes what is written in the quotation. This belief should not be confused with what is claimed in the title, as a "proof" of anything. While this is a "proof" according to what the author believes in and interprets, it does not follow that his interpretation is correct. It is certainly not a proof of what the author says next about the "place he went." Which was "real," later he would say more real than real. Here, he only says that the life we are living "here and now" is like a dream compared to the reality of his experience. Since the author *did not go* to any place – therefore, the "no-place" is not real!

The rest of the prologue is intended to convince the reader that the author is qualified to examine and analyze his experience. He has experience in medical analysis, and he is familiar with "the most advanced concepts in brain science and consciousness studies." Therefore, the readers should trust what he is telling them. Not only that, but the author claims that his journey was true, and therefore he felt he must tell it to others. That was the chief task of his life.

The author clearly confuses (or perhaps cannot tell the difference between) the experience he had, with the interpretation of that experience. I believe that the author's experience of NDE was true. However, it is also clear that his interpretation of that experience is anything but true!

Finally, the author concludes the prologue with the bombastic, pretentious statement:

"What I have to tell you is as important as anything anyone will ever tell you, and it's true."

This statement brandishes a red flag to the reader and should dissuade him/her from reading this book. Unfortunately, many readers have fallen into this trap.

The only reason why I chose to continue reading this book is because I wanted to write a fair review and to this end, I had to finish reading it so that I do not run the risk of misinterpreting the prologue.

I will go through the chapters in the following pages, commenting briefly on what the author tells us. It will be brief because there is nothing significant, interesting, important, or convincing in what the author has to say in the entire book. I personally did not like the manner in which the author narrated his story, switching from the real world, i.e. what transpired in the hospital and to his family, and what happened to his dream (or, as he puts it, his journey to heaven). The blurring of the lines, reality straddling dreamland does not, in my opinion, contribute in convincing the reader of the reality of his experience. Reading the whole book made me highly skeptical as to the author's credentials as a scientist.

5. Chapter 1: The pain

This chapter consists of some biographical notes, focusing on his meeting and marrying his wife Holley, and his pain on November 10, 2008. Then we are told that during seven days he would be "present" to Holley and the rest of his family only in body. He did not remember anything of this world during that week, he was unconscious.

"My mind, my spirit – whatever you may choose to call the central, human part of me – was gone."

We have read this description before, when he said that "his brain," the part that makes us human was absent. Now, it is the central human part of him that was gone. Absent? Gone? Where to? The rest of the book is supposed to answer these questions.

6. Chapter 2:   The Hospital

In this chapter, we become silent witnesses to the author's plight as he allows us to take a peek into that hospital room when he was diagnosed with E. coli meningitis, or perhaps a brain infection. We are informed about his diagnosis; perhaps E. coli meningitis, or perhaps brain infection. Nothing relevant to book's main story. Of course, nothing here can be criticized. That is the story that actually happened. The author sets the stage, the real stage for his dramatic sojourn.

7. Chapter 3:   Out of nowhere

The real drama begins here:

"Then, out of nowhere, I shouted three words…"

These words were heard by everyone who was present:

"God, help me."

Are these three words coming from the "real world" or from the "other world"? After uttering those three words, everyone rushed to the stretcher and by the time they got to him, he was completely unresponsive.

For those who have read articles, watched movies with the same theme/experience as his "shouting," (I have read and watched similar experiences) the person is supposed to be aware of what is happening to him, his mind is active, his eyes can see, and although in his mind he is saying something, it's just it, only in his mind, because no one knows, not the doctors, the nurses or whoever is close to patient knows he is crying out to be saved. In short, it is clear that he *really* did not die, and was not *really* beyond the grave.

8. Chapter 4: Eben IV

Eben IV is the author's son who was summoned to the hospital while his father was in the intensive care unit (ICU). The son gets to the room, and a poignant scene unfolds. His son froze in the doorway when he saw his father. His son looked at him as if he was looking at a corpse. Of course, his father's body was there, in from of him, but "he was gone," adding:

"Or perhaps a better word to use it: elsewhere."

I do not like the use of the term corpse in this story. If Dr. Alexander had *gone, it must have been elsewhere.* This placed called "elsewhere" is described in the next chapter.

9. Chapter 5: Underworld

The author writes here an impressive description of a place called "elsewhere," in the previous chapter, which he now refers to as the "Underworld."

I must admit that the prose, almost poetic, tugged at my heartstrings. Here are some highlights:

"Darkness, but a visible darkness."

"Transparent, but in a bleary, blurry, claustrophobic suffocating kind of way."

"Consciousness, but consciousness without memory or identity."

These are very beautiful words. At this point the author is so engulfed with these beautiful words that he probably forgot the title of the book; having no idea where he was, and who he was, do not constitute a "proof" that the Dr. *Alexander* was in heaven. Here is some more of the same:

"Sound, too: a deep, rhythmic pounding, distant yet strong, so that each pulse of it goes right through you. Like a heartbeat?"

He had no idea how long he was in that world, in that place, where there was no "sense of time." He had a feeling that he had always been there and would always continue to be there.

This, and other surrealistic descriptions of the netherworld does not leave any doubt that everything was a dream, rather than having visited a new, "real" world.

He did not feel like a human while I was in "this place." Not even an animal.

"I was simply a lone point of awareness in a timeless red-brown sea.

Then, his sense of deep, timeless, and boundary-less... gave way to a feeling like he was not:

"really part of this subterranean world at all, but trapped in it."

So far it is a nice description of one who dreams, but technically he was in a coma so it is not clear to him or to anyone else what the real situation was.

Then, he was jolted back to something real; a smell a little like feces, like blood and like vomit. Describing this smell as:

"...of biological death, not of biological life."

It is not clear where the smell came from, the *real world*, or from the *other world*.

Towards the end of the chapter we are told that something "new emerged from the darkness." That something was not cold, dead or dark, but the exact opposite of all those things. He was so overwhelmed, by that "something," that he could not describe "how beautiful it was."

"But, I'm going to try."

We are also totally in the dark what it was that emerged from the darkness. He will tell us this, albeit not immediately, but again he goes on to say something that went on in the real world, presumably while he had experienced difficulty in describing this beautiful thing.

10. Chapter 6: An anchor to life

We are back at the hospital, family members visiting the author, and worrying about his condition. Then, some thoughts about the value of one's family: "Your family is who you are." Nicely crafted, but is it really true?

Next comes some more biographical notes about his studies, his belief in God, heaven and afterlife. He is correct in claiming that science seems to "push our significance in the universe" closer and closer to zero.

"Belief would have been nice. But science is not concerned with what would be nice. It's concerned with what is."

It is true that science is not concerned with what is "nice," but it is not true that science pushes our significance in the universe closer to zero!

He then talks about science which is incapable of dealing with consciousness and other phenomena of life. Again, since he was a scientist he knew what the "brain really is," nothing more than a machine that produces the phenomenon of consciousness. Of course, science leaves very little room for "the soul and the spirit," or for the continual existence of a personality after the brain that supported it stopped functioning.

Close to the chapter's end, the author mentions the "absolute honesty and cleanness of science," and the fact that he respected it as it left no room for "fantasy or sloppy thinking." He continues by saying "if a fact could be established as tangible and trustworthy, it was accepted. If not, then it was rejected." Interestingly, the author mentions that the acceptance of a fact depends on its tangibility and trustworthiness. How then can he claim he has a "proof of heaven" when he alone experienced what he did, which no one can refute? How does he address the issue of tangibility when he talks about heaven which is, needless to say, an abstract concept?

It seems to me that the author is paving the way for the reader to accept the fact that science cannot deal, nor explain all those religious concepts, and in particular "afterlife," or "life everlasting." Yet, at the same time he uses his scientific background to lend credibility to his story.

11. Chapter 7: The spinning melody and the gateway

This chapter brings us back to the other world. Here is the beautiful introduction:

"Something had appeared in the darkness.

Turning slowly, it radiated fine filaments of white-gold light, and as it did the darkness around me began to splinter and break apart."

Then he hears a new sound: a "living sound," like a beautiful piece of music. Then, a "pure light" descends, the light

"got closer and closer…generating those filaments of pure white light that…were tinged…with hints of gold."

All these fantastic and vivid descriptions of what he has seen and heard, seems to "prove" that the author was either dreaming, or hallucinating. Nothing real!

He felt like he was being "born," in an incredible beautiful world. He was also sure it was not a dream.

"Though I did not know where I was…, I was absolutely sure of one thing: this place I'd suddenly found myself in was completely real."

This is the ultimate *proof* that what he had experienced was a "beautiful, incredible dream world…" What is really incredible is that the very same dream is not a dream – but this is real!   The author is clearly unable to distinguish between a dream and reality!

It is unfortunate that the author, who recognized his experience as a dream, claims without any evidence, let alone proof, that what he had experienced was *"completely real,"* and that this fictitious reality was a "proof of heaven."

If Alexander were not a scientist, he could be forgiven for confusing a dream with reality and using that confusion as a "proof of heaven." The fact that he is a neurosurgeon renders such statements unforgivable, if not outright deceptive.

It seems that the author is aware that the word "real" is abstract and perhaps undefinable, and yet he uses the word adding adjectives such as "completely real" or "true," or something even "more real than real." Remember that he did not have a sense of time, and he could not tell how long he was flying there. Then he says something I did not understand that "Time" in that place was not the "simple linear time we experience on earth."

I have no clue as to what he means by "linear time." However, he had a previous statement that time was timeless in that world. Perhaps, the meaning of this word is also meaningless.

In the next two pages, the author describes his beautiful experience. First, he felt that he was not "alone up there." Then we find a description of his encounter with an equally beautiful girl:

"She was wearing the same kind of peasant-like clothes that the people in the village down below wore. Golden-brown tresses framed her lovely face."

Where exactly is the village "down below?" Does he mean in the real world, or in the underworld?

"The girl's outfit was simple, but its colors… had the same overwhelming super-vivid aliveness."

A really beautiful surrealistic description of a surrealistic world.

The girl looks at him. "It was not a romantic look. It was not a look of friendship. It was a look that way beyond all these…"

I can see clearly why the look was indescribable in simple words. Simply stated, the encounter was imaginary, and a figment of the author's hallucinatory state, and as such no words could have described it. Or perhaps, it was a romantic look and he might have responded with a romantic look, but he shied away from telling the story?

Next, we are led from one fantasy to another. She talked to him, but without using words. Yet, he got the message she conveyed to him. The message went through him "like a wind." A message he understood instantly that "it was true."

I wonder, if the experience were indeed true, where is that woman now?

Her message remains a secret because she spoke with no words, and yet surprisingly, the author "instantly understood that it was true."

At this point, I would imagine any rational reader to put down the book and stop reading it. My reason for continuing to read the book in its entirety was to see how far he would go on in describing his experience as both "real" and "true."

That girl, or perhaps I should say that Girl, though she talked without words, communicated to the author a message he got written in simple words. The author tells us that he has *translated* the "no words" message into the "earthly language."

The following is a translation of the message:

"You are loved and cherished, dearly, forever."

"You have nothing to fear."

"There is nothing you can do wrong."

What a beautiful message from a beautiful girl who spoke without any words, and yet, he could translate her uncommunicated words into an earthly message. Then she promises to show him many beautiful things – again without using words, but he could still understand the message. Finally, she tells him that:

"eventually, you will go back."

"Back where?" the author asks the girl, and I am wondering whether he used "earthly language," or the "no-words" language to ask her this question. All of these sound like a child's dream, or make-believe world.

The chapter concludes with the most unconvincing statement. That he is a scientist, and that he knows his biology, and he can distinguish between "fantasy and reality."

I doubt it very much if the author knows the difference between "fantasy and reality." He repeatedly pounds on this in order to drive home the point to his readers with the hope of convincing them of his fantastic story. If that was the "single most real experience" of his life, then it is clear that all his life was one continuous dream.

I should also add here that whatever the author had experienced in this "other world" he must have been lucky. Other people with similar experience were not so lucky, and they heard or might have heard more unpleasant words. I will comment on this later.

12. Chapter 8:   Israel

In this chapter the author jumps back into the real world and narrates how he saw his family support him in his dire condition. It is no accident that stories about the "real world" are interspersed with stories about the "other world." The author seemingly wants to convince the readers of the realness of both worlds. He also talks about how he must have contracted his illness.

13. Chapter 9:   The core

From the world of reality, he switches back to the other world where he found himself in "a place of clouds." He then describes his difficulty in writing about his experience, using simple words used in this world. And as for "the beings themselves," they were different from anything that he had known "on this planet." ***They were more advanced. Higher.***

"More advanced," "higher" are meaningless adjectives in this context, where he could not express anything in words.

Then, he hears a sound, "huge and blooming," which came down "from above."

 "The sound was palpable and almost material, like a rain that you can feel on your skin but that doesn't get you wet."

This device is like a "warm wind," a "divine breeze," and the author asks some questions to this wind, which transformed into a divine being:

"Where is this place?"

"Who am I?"

"Why am I here?"

He immediately got the answers (In word? In which language?).

"Each time I silently posed one of these questions, the answer came instantly in an explosion of light, color, love, and beauty that blew through me like a crashing wave."

This is really a beautiful description of nothingness.

Then, he sees a new light:

"...a light that seemed to come from a brilliant orb that I now sensed near me. An orb that was living and almost solid, as the songs of the angels had been."

He felt like "a fetus in a womb," and in this case, the "mother" was "God, the Creator, the Source who is responsible for making the universe and all in it."

"This Being was so close that there seemed to be no distance at all between God and myself."

So, he exchanged questions and answers with God, the Creator Himself! That Being is described as being "warm" and "personal" and that this being possesses qualities that we humans possess, but only in "infinitely greater measure."

These beautifully crafted yet meaningless words go on, and on until the end of the chapter. He repeats the same argument that he spent all his life accumulating knowledge "the old fashion way." and now the new and "advance level of learning" gives him food for thoughts for ages to come.

Remember that this chapter is about the Core, (chapter 12 is also The Core) and the Core is the "wind" that transfigured into God himself, with whom the author had conversed with, sans any exchange of words between them. I don't think that even Moses himself had that great privilege of seeing and talking to God, up close and personal. If I had experienced the same thing, the first thing I would have done was to ask that "being" for his/her ID card in order to verify his/her identity.

14. Chapter 10: What counts

In this chapter we are back to the real world. The author reveals that he is an adoptee, and that no matter how secure and loved he always was by his adoptive parents, he had always nurtured the idea of meeting his biological parents someday. That encounter was not to be, as his biological parents rejected contact. The chapter also tells us that he had lost faith in the existence of a Being as a result of that painful rejection. Poignant, and touching as this chapter might be, I do not see its relevance to his "proof" of heaven.

15. Chapter 11: An end to the downward spiral

This chapter deals with the author's epiphany – finally meeting his biological parents. He also talks about finally shedding of the notion that for a long time he had not been loved and believed that he did not deserve to be loved.

Touching and riveting, no doubt, but I still do not see its relevance, nor its' lending any credibility to his "proof" of heaven. Perhaps, this feeling of lack of love was the reason for a dream with overflowing love?

16. Chapter 12: The Core

Another one of those well-written narratives of what the author "witnessed." He puts more weight however on his depiction of his wonderful journey, and less on the real world. He purposefully melds reality with his journey to the other world, with the purpose of making the latter real.

The unwary reader might fall into the trap of blurring the lines between the real and the unreal, thus accepting them both as true. This melding effect has an opposite effect on me. I tend to believe that not only are his dreams unreal, but the author himself, in the *real world* is making un-real judgements and conclusions.

Notwithstanding my objection to the author's shuttling back and forth on his narrative between real life, and the "other" life, I believe that the author's description is honest. It is his interpretation of that experience that I have problems with. My conclusion is reinforced by the fact that he hammers on his profession, that of a neurosurgeon, and that he previously did not believe in NDE stories but had a change of heart when he himself "experienced" and analyzed it. His proof; he analyzed his situation in "scientific standards," and as a neurosurgeon he knew what he was talking about, and that very same experience is a "Proof of Heaven."

Beautiful non-verbal exchanges between the author and his companion is narrated in a vivid, almost pictorial style. His companion was there all the time, but now she was once again "in human form." Again, she promised to show him many things, but eventually he will be going back. It is only now that he understood what was meant by "going back." This also led him to the understanding that he did not belong to "this place," but he was only visiting it.

Once again, we find the same message that he got before (You are loved and cherished, you have nothing to fear and there is nothing you can do wrong.)
My question is, in which language?

As if this was not clear enough, the author explains the message he got, in one sentence and in plain words:

"You are loved."

This is certainly a beautiful, moving experience, but totally unconvincing as to its reality. The author makes use of flowery and powerful language in order to convince the reader to believe that his experience was "the reality of realities," the "glorious truths," and the "core of everything that exists."

Since I do not know what the "reality of realities" is, and I suspect that also the author does not know that, I can only conclude that the "reality of realities" is another expression for "unreality."

I would add to this that the whole story is the "fantasy of fantasies," "the mother of all dreams and imagination."

Anticipating resistance from the reader, the author starts by asking a rhetoric question to the reader: "Not much of a scientific insight?" which follows by his answer: "Well, I beg to differ."

Once again he brandishes his credentials as a neurosurgeon, and claims "that this is also the single *most scientific truth…* " He banks on his being a scientist, a doctor, in order to say to the readers that they have to believe anything that comes from him. He wants the readers to believe what he peddles as "scientific truth." I warn the readers not to be deceived with all the author's sweet talk, for he is, strictly speaking *not* a scientist, *does not* think like one, and does not know what scientific truth means. If this experiment is the "most important scientific truth," then I have no choice but to conclude that his scientific experience is untrue!

From the "other world, NDE, to going back to love…the perfect ingredients for a mishmash of a chapter. He also talks about unconditional love as something "banded around a lot…," and asks "how many people can grasp what unconditional love truly means." To this I say, one does not need an NDE or a degree to experience unconditional love. We see it daily; we see it in regular, common people. We see parents of autistic children who give all their best, providing their special children with all the love they can which might never be reciprocated with the same degree. We see men and women who look after their Alzheimer stricken partners, giving them all their love, all their best, knowing that at some point their partners cannot even remember who they are.

I will skip Chapter 13, which does not contribute anything relevant to the proof of heaven and proceed to the next chapter.

17. Chapter 14:   A special kind of NDE

It makes me wonder why the title is such, could it be that because the NDE happened to a neurosurgeon who knows what he is talking about? Could it have been equally special if it happened to Tom, Dick, or Harry?

The wordsmith is at work even harder, as the author attempts to floor his readers by saying incredulous statements. After that experience he understood that he "was part of the Divine," and that nothing could ever take that away.

This kind of description does not leave much doubt about the "Scientific" experience he was going through. And toward the end of the chapter, we read:

"At the risk of oversimplifying, I was allowed to die harder, and travel deeper, than almost all NDE subject before me."

He died? Really? He died, and came back to life, just like the story of the resurrection. The story gets weirder, and weirder, and without failing to mention that he died harder than most NDE subjects. I wonder if this is also the case because he happens to be a neurosurgeon…

18. Chapter 15:   The Gift of Forgetting

For the very first time, the author discusses the possibility that all our perceptions, his experience not exempt, are merely illusions:

"But our perceptions are just a model – not reality itself. An illusion."

He backtracks after discussing the possibility of "an illusion." Perhaps he had a "slip of the tongue" episode, and wanted to rectify his mistake, so he goes on to ask himself "Why am I so sure of all this?"

His answer is far from being convincing: First, he says, because he "was shown it" and second, because he "actually experienced it."

I am sure the readers are as confused as I am as the author sends us mixed signals. Or perhaps, the author himself is confused, as simple as that!

This is another statement I would not expect to hear from a scientist.

And then, he continues to tell us further unfounded ideas or views that even on earth there is more good than evil, but it will be impossible for evil to gain influence "at higher levels of existence."

I have no idea how the author knows that this is "entirely impossible" at the higher level of existence, which is nothing but a fictitious level of existence!

From one illogical statement to another, he waxes spiritual and claims that one of the biggest mistakes people make, about God, is to imagine "God as impersonal."

No comment! Personally, I never did that particular "biggest mistake" so I am happy.

Chapters 16 to 19 contains nothing relevant to his NDE.

19. Chapter 20: The Closing

More highly incredulous statements that left me scratching my head:

"In the presence of my guardian angel on the butterfly's wing and an eternity learning lessons from the Creator and the Orb of light deep in the Core."

Again, one *kiskush* follows another *kishkush*.

The next chapters are more of the same, illogical, fantastic, figments of the author's imagination. The author is transformed from a non-believer to someone who swears by what he says. Then again he asks a rhetoric question:

"Who wouldn't be anxious to hear of my discoveries?

His answer: Quite a few people, specifically, people with medical degrees.

Here is a hint for the reader. He did not believe that entire story – *before,* but now he knows the "truth." Much of the rest of the book demonstrates how the author persuades his readers that he, a scientist who understands the brain knows what he is talking about. In some instances, he expresses contempt to those who are not smart enough to accept his message as the ultimate truth, more true than true in fact, and more real than reality itself. Perhaps he is also trying to solicit his readers' unconditional acceptance of his message as Bible truth, or perhaps feel envious that he had experienced a *direct* communication with God.

"Communicating with God is the most extraordinary experience imaginable, yet at the same time it's the most natural one of all, because God is present in us at all times. Omniscient, omnipotent, personal – and loving us without conditions."

This sounds more like a priest standing behind the lectern preaching, rather than a "scientist" narrating his experience. It also happens to be nothing new to believers; they have heard it so many times.

As the book winds down, he reminds us again that he is a scientist, a doctor. Hence, as a result of his duty to honor the truth and to heal, he has also the duty to tell his story.

"A story that as time passes I feel certain happened for a reason."

If the author wants to "honor the truth" he should warn the reader that his experience is anything but true.

His recapitulation:

"I'm living proof."

Proof? Of what? If Dr. Alexander were a real, honest-to-goodness scientist, he should know what *proof* means in science. I would like to believe that I read the book with eagle eyed preciseness, and nowhere did I find any proof of anything, and certainly not a "proof of heaven," nor the vaguest evidence of its existence. The fact that a scientist uses the word *proof* as the title, as well as in countless places in the book is a "living proof" (whatever this means) that this book is replete with anything but a *proof of heaven*!

20. Some final remarks on possible consequences of reading this book and a kind suggestion to the author.

I believe that Alexander' story is true. I also believe that he firmly believes in his *interpretation* of that experience. His narration of his NDE, written in flowery language is a testament to his glorious encounter with angels, seeing Heaven, and meeting God himself. However, I beg to differ with the author's interpretation. If I were to have an NDE, I would not claim that the experience, even remotely, a *proof* of *anything*. I certainly would not even think of writing a book on it.

The frequent use of the words "proof," "real," and "truth" throughout the book is a "living proof" that the author knows that his story is – well, just another NDE story. It is inconceivable that such book was written by a scientist. No one, including the best scientists, in my opinion have a right to claim that his/her NDE (or dream, or hallucination) is "real" or "true."

As a final comment, I want to remind the reader of what I wrote before. The author was lucky to have experienced an NDE with so much beauty, love, caring, and reassurance.

Which brings me to the question:

What if?

What if another person experiences NDE but is not as lucky, and instead of love and goodness sees ugliness, seething hatred, and evil? One outcome of such an experience is to write a book on the "Proof of Hell," and make some money out of it. However, further beyond this possibility is are even more detrimental consequences.

Reading such a book might be detrimental to others. People who are treading a fine line between reality and the un-real, and who might have experienced similar visits, and have heard voices might process the "message" differently, and resort to destructive responses.

From Biblical times, passed on from generation to generation, stories have found their way into our present consciousness. One such story was Abraham's willingness to slay his beloved son because he heard a voice from heaven commanding him to do the ultimate sacrifice in order to prove his loyalty to God. This story about man's complete and blind surrender is taught to children at an early age, at the time when the brain's suppleness allows impressions and visions to be ingrained in a young child's brain.

For someone like me who hails from the Middle East, the story of extremist radicals who have fomented evil deeds could not be any closer to home. The stories may vary, but they somehow share a common denominator, people's readiness and willingness to answer the call from a voice that either directly, or indirectly tells them what to do.

My apprehension stems from the fact that someday, someone might have a similar NDE as narrated in Alexander's book, and take it as kosher-stamped, *real* and *genuine*, thus affording him/her the liberty to do anything that was dictated to him/her.

As I am writing this paragraph I heard the news that a man beheaded his wife. The police asked him why he did that. His answer: "I heard a voice telling me to do so."

This is not a direct criticism of Alexander's book, but rather a red flag that should warn people of the way they should interpret a book such as the author's.

I therefore suggest to the author to be more careful, and responsible, and perhaps drop the words *true*, *reality* and *proofs* in the next edition of his book, and to warn the readers not to take his story too seriously.

I do not have any illusions that such a correction of this particular book will have much effect on hundreds, if not thousands of individuals with twisted views to begin with. Assuming that they will have diametrically different NDEs compared to the author's, one of evil, hatred, and ugliness, I cringe just thinking about the consequences. I trust though that there are more intelligent, sensible readers who might read the next edition of Alexander's book and will be discerning enough to either accept or reject the author's claims.

Notes

Note 1: Probability and conditional probability

There is no "definition" to the word "probability". If you look at any book on Probability you will realize that the definition of the term "probability" uses the concept of probability. In other words, these definitions are circular, i.e. the definition of probability is based on the concept of probability or another equivalent concept such as randomness, likelihood etc. Some books introduce the concept of probability *axiomatically* which means that a number is attached to every *event*, and this number fulfills some properties.

I suggest that you read the first paragraph in Wikipedia on "Probability." You will find the following (as of Jan 2013 – this might change with time, as Wikipedia's contents evolve with time).

Probability *is a measure or estimation of how likely it is that something will happen or that a statement is true. Probabilities are given a value between 0 (0% chance, or will certainly not happen) and 1 (100% chance, or will certainly happen). The higher the probability, the more likely the event is to happen, or, in a longer series of samples, the greater the number of times such event is expected to happen.*

From an encyclopedia: *Probability Definition*

1. The quality or state of being probable; appearance of reality or truth; reasonable ground of presumption; likelihood.

2. That which is or appears probable; anything that has the appearance of reality or truth.

3. The chance that a given event will occur.

4. Likelihood of the occurrence of any event in the doctrine of chances, or the ratio of the number of favorable chances to the whole number of chances, favorable and unfavorable.

In this Note, I hope to convince you that in spite of the fact that there exists no definition of probability you already know what probability is. You have a sense of what it means when one says that the probability of an event A is larger, or smaller than the probability of an event B.

This sense-of-probability is much the same as the sense-of-beauty, or a sense-of-time, you have without having a definition of "beauty," or of "time." You can confidently say that one landscape is more beautiful than the other. Similarly, you can tell that one event was longer than the other. These examples should not mislead you into concluding that the sense-of-probability is always a subjective sense. While it may be subjective in some cases, it is an objective measure in most cases when it is used in the sciences – objective in the sense that everyone who uses these measures will agree with them.

Read the following two statements and assess their objectivity or subjectivity.

(i)   If you marry this lady it is very probable that you will be happy all your life.

(ii)   If you bet on the outcome {"4"} in the throwing of a fair dice, it is probable that you will win on average in one out of six throws of the dice.

Which of these probabilities is more subjective or objective?

I hope that you are comfortable with the intuitive meaning of probability. I also hope that you gave the correct answer to the question posed above. The answers you gave were based on intuition, or on your sense-of-probability.

(1) The classical "definition" of probability

Note that I enclosed the word "definition" in quotation marks. You will understand why, soon. For the moment let us "invent" or "discover" this definition by ourselves.

Example: We throw a fair dice in such a way that each outcome has the same likelihood of occurrence. What are the probabilities of the following events?

(a) The outcome is {4}

(b) The outcome is an even number, i.e. it is one of the results {2} or {4} or {6}. We write this event as {2,4,6}

(c) The outcome is greater than 4. This means it is either {5} or {6}. We write this event as {5,6}

The answers are: (a); 1/6, (b); ½ and (c); 2/6.

How did we assign probabilities to these events? We assumed that the dice is "fair," and that we threw it in such a way that each single possible outcome has the same likelihood. This means that the dice is a perfect cube, and its mass density is evenly distributed, and that we threw the dice in such a way that it rolls and spins in the air many times so that it will eventually land on the floor with one of its sides facing upwards.

Before we calculate the probability of an event we have to decide on the *range* of values that probabilities can attain. The choice of the range is arbitrary. This is similar to the choice of scale for the temperature. The most common choice of range is between zero and one (see also next session). We often use the scale between 0% and 100%. The lowest probability value, zero, is assigned to the *impossible* event. The highest probability value, one, is assigned to the *certain* event.

Now for the event (a) in the example above, we reason that since there are altogether six possible outcomes, and we assume that each outcome has the same likelihood of occurrence, therefore, the probability of a single outcome, say, {4}, in case (a) is 1/6.

If you pause and think about the reasoning that led us to the number 1/6, you will find that we used the phrase "each of the outcomes has the same likelihood" which is tantamount to saying that each event has the same probability. Once we also fixed the value of the *certain* event to be one, we can calculate the probability of each single event as being 1/6. Thus, what we have done is not to *define* the probability of the event {4}, but to assume that we *know* the probability of the event to be 1/6. In other words, this "definition" is *circular*. It uses the concept of probability (or likelihood or chances or odds) to "define" the probability. Sometimes an argument based on symmetry, or equivalence of all possible results is used to reach the conclusion that the probability of each event is 1/6.

Let us go to case (b). What is the probability of the occurrence of the event {2,4,6}, i.e. the outcome is an *even* number?

There are two ways of reasoning. First, we can argue that there are altogether six equally likely outcomes, each having probability of occurrence 1/6. Therefore, the occurrence of either {2} or {4} or {6} must be larger than 1/6, and most likely to be the sum of these three probabilities, i.e.: $\frac{1}{6} + \frac{1}{6} + \frac{1}{6} = \frac{3}{6} = \frac{1}{2}$.

The second reasoning is to divide all possible outcomes into two groups of events; "even" and "odd" outcome. Think of coloring all the faces of the dice with even number of dots with red, and all the faces of the dice having an odd number of dots with blue. The probability of the event "even" is equivalent to the probability of the event "red" face. Since there are two possible outcomes either "red" or "blue" and since we believe that the dice is fair, we conclude that the probability of the event "red" (or "even") is ½.

Note again that in calculating the probability of the "event" we used *probabilistic* arguments, i.e. we assumed that each outcome has the same likelihood. Therefore, this method of calculation cannot be viewed as a *bona-fide* definition of the probability of the event "even."

Let us turn to the case (c). The event "greater than 4" means that the outcome is either {5} or {6}. We write this event as {5,6}. Using the same type of argument as before we can conclude that the probability of the event

{5,6} is the sum of the probabilities of the single outcomes {5} and {6}, i.e.
$\frac{1}{6} + \frac{1}{6} = \frac{2}{6} = \frac{1}{3}$.

We see that the classical "definition" is intuitively clear. You should realize however that this is not a *bona-fide definition* of probability. The "classical definition" already assumes that we *know* the *probabilities* of each single outcome. Therefore, this definition is circular.

Furthermore, this rule of calculating probabilities does not apply in general. First, it is not always clear what the *elementary outcomes* are. We shall see examples of such cases in the next session. For now, it is sufficient to say that in the case of throwing a dice we assume that there are six possible outcomes (we neglect the possibilities that the dice will fall on an edge or on a vertex, or perhaps it will fall and break into pieces so that no clear outcome is observed). The most important aspect of these probabilities is that we all agree on them. This is not the case when we ask for the probability that "it will rain tomorrow" or that "this book was written by Shakespeare." It is certainly not the case when someone talks about the probability of God, or the probability of Kukuriku.

(2) The relative frequency "definition" of probability

In all of the examples discussed above, we started with the assumption that there exists a finite number of elementary events, or elementary outcomes, and that these have equal likelihood (or chances, or probability) or occurrence. How do we calculate probabilities in cases where there are no obvious elementary events? How are the probabilities of such events defined? The general and honest answer is that there is no *bona-fide* definition of probability, nor a method of calculation which is satisfactory for the general events.

What is the probability that a dormant volcano will erupt in the next hour? What is the probability that tomorrow the sun will explode? What is the probability that a child will be born with four legs, six arms and three heads?

Clearly, there is no way of defining, let alone calculating the probabilities of these events. Yet, people do use the term probability in connection with such events. The only meaning that "probability" has, in such context, is the extent of one's belief on the chances of occurrence of that event.

However, there is a large class of events for which one can offer an "experimental" way of calculating their probabilities. These are the cases when we can repeat an experiment many times, or certain events have occurred many times in the past, and we can collect "statistics" on specific events. For example, suppose we have a dice which is known to be unfair, say a dice with an asymmetric distribution of mass, or a partially broken or twisted dice, we obviously cannot assume that each outcome has the same probability.

In this particular example, we apply the so-called *relative frequency* "definition" of probability.

We throw the dice many, many times say a thousand times, and collect the "statistics" about the frequency of observing the result {1}, {2}, {3}, ···

{6}. By "frequency," we mean the ratio of the number of times as specific result occurred and the total number of throws.

Suppose we found that after a thousand throws the following results:

50 results showing {1}

100 results showing {2}

100 results showing {3}

200 results showing {4}

250 results showing {5}

300 results showing {6}

We might tentatively *assume* that the probabilities of the different outcomes are

$$\frac{50}{1000} \ , \quad \frac{100}{1000} \ , \quad \frac{100}{1000} \ , \quad \frac{200}{1000} \ , \quad \frac{250}{1000} \ , \quad \frac{300}{1000}$$

The relative frequency definition states, that if we throw the die *infinite* times, the fraction of times each outcome occurs is the probability of that event.

This definition is problematic at best. In fact, it is also circular in the following sense. We believe that if we do the experiment infinite times the fraction of times each outcome will occur will tend to some constant value between zero and one. Unfortunately, we cannot perform an infinite number of experiments for many experiments. In fact, no one can guarantee that if we calculate these fractions for many experiments the fractions will tend to some constant values.

In practice, we take a large but finite number of experiments, collect the "statistics," as we did above and *assume* that these results are the approximate probabilities of the outcomes. We believe that if we repeat the experiment many times (thousands, millions, billions…) it is *highly probable* that the fractions we get are the "true" probabilities. You see that we use the concept of "probable" to define the concept of probability. Therefore, this method cannot be considered as a *bona-fide* definition.

Yet, in practice we use this method with finite number of experiments to estimate the most likely probabilities. It is not perfect, it does not guarantee that we will get the correct results, and it is not always applicable. Yet, *this is what we have*, and in many cases this method is very useful.

Based on these methods we can determine the approximate probabilities of outcomes of an unfair dice as we did before. Doctors and pharmaceutical companies determine the efficacy of certain drugs. Insurance companies estimate the likelihood that a certain person (within a specific age bracket, sex, education, marital status, etc.) will be involved in an accident, and with this estimate they will calculate the cost of one's insurance policy. In all of these cases, and in many others we do not have "exact" probabilities, but this is the best we have, and we use them because we *need* to use them.

You might wonder how this method compares with the classical "definition." The latter sounds more accurate, more precise, more reliable, but this is only an illusion.

First, we can never be sure that the dice is perfectly fair (whatever that means). If we are not sure, we must use the experimental method, and if we find that each outcome occurs with the same frequency, i.e. about $\frac{100}{600}$, or

$\frac{1000}{6000}$, etc. then we can be reasonably sure that it is a fair dice, and that the probabilities are 1/6. But what if we are (somehow) sure that the dice is fair, how do we know that the probabilities of the outcome are equal to 1/6? In fact, we do not *know*. We believe that this is a reasonable assumption. If we doubt this assumption we can do the experiment, or we can imagine making the experiment many times. Of course, in our *imagination* we can afford to repeat the same experiment *infinite* number of times, and *imagine* that the relative frequencies will all be equal to 1/6.

(3)   The axiomatic approach to probability

As in any other branch of mathematics one starts with a few axioms. These are essentially postulates on which everyone agrees. On these postulates one builds up the entire theory. It should be noted that probability theory is relatively a "newborn" to mathematics. It was only in the 1930s when the axiomatic theory of probability was established. Here, we only mention the so-called axiomatic approach to probability. Some authors refer to this approach as a *definition* of probability. It is not! This approach does not *define* probability, but only assumes that there is a quantity, which we call probability, which satisfies some requirements. For details, see Ben-Naim (2015b).

(4)   Independence and dependence between events

One of the most important concepts in the theory of probability is the *conditional probability*. This concept is central to the theory of probability and does not feature in set theory.

As a quick warm-up, consider the following easy problems:

(a) You are shown two urns. In each of the urns there are two red marbles and two blue marbles. What is the probability of picking a red marble from the left urn?

What is the probability of picking a red marble from the right urn?

What is the probability of picking a red marble from the left urn *and* a red marble from the right urn?

(b) You are shown one urn containing two red and two blue marbles.

What is the probability of picking a red marble?

Suppose you pick a red marble in the first trial. You return the marble to the urn, shake the urn, and try once again to pick a marble.

What is the probability of picking a red marble on the second trial?

(c) As in (b), you start with the same urn, having two red and two blue marbles.

You pick one red marble. The probability of this event is ½.

Now suppose you pick a red marble in the first trial. You *do not* return the marble to the urn. What is the probability of picking a red marble in the second trial given that you have already drawn a red marble in the first trial, but the marble was not returned to the urn?

If you answered the questions in (a), (b) and (c) correctly you already have a sense of what independence, or dependence between events mean. Not only a sense, but you also have correctly calculated a *conditional probability*.

In general, two events are considered to be *independent* when the occurrence of one event has no effect on the probability of occurrence of the other event. For instance, throwing two dice at two different places are independent experiments, and their outcomes are independent events. On the other hand, if the two dice are tied by some inflexible wire, the outcome on one dice affects the probability of the outcome on the second dice.

Similarly, if you draw a marble from two urns the outcomes are independent. The same is true if you draw a marble from an urn, then return the marble to the urn, and draw again. The two consecutive experiments are independent.

On the other hand, if you draw a marble, see the result, and do not return it back into the urn, clearly the probabilities of the outcomes on the second draw would depend on the outcome of the first draw. Thus, in example (c) above, *given* that you drew a red marble, the probability of drawing a red marble is 1/3, and a blue is 2/3.

The concept of dependence is also clear for events which we encounter in daily life. Suppose that it is raining in Jerusalem today. This fact has (almost) no effect on the probability of raining the next day in New York. However, if it is raining in Jerusalem today, then, it might affect the probability that it will rain in Jerusalem the next day. This is of course a very intuitive and qualitative assessment. We shall be interested in a more precise definition and methods of calculating probabilities of dependent events.

Suppose I throw a dice and I tell you that the result is "even." What is the probability that the result is {4}? Note here that you have given some *information* on the result. It is not the answer to the question we asked, but it is a "hint" that might help us find the answer. We write required probability as

$$P("4"|given\ that\ the\ result\ is\ "even")$$

or for short

$$P("4"|\ "even")$$

In this particular example you should be able to calculate this *conditional* probability, presuming that the dice is fair, and that you know that the result is "even." This means that the outcome could be either {2}, {4} or {6}. Each of these has the same probability; *one* out of *three*. Hence, the answer to the question is 1/3.

Note that when I ask you what the probability of the outcome {4} is, giving you only the information that the dice is fair, the required probability can also be written as a conditional probability:

$$P("4"|given\ that\ the\ one\ of\ the\ results\ between\ 1\ and\ 6\ occurred)$$

We usually refer to the last probability as simply the *probability* of {4}, omitting in the notation that one of the results between 1 and 6 occurred.

Now that you have an intuitive feeling on what independence and dependence between events mean, let us try to find the general rule of calculating the probability of occurrence of both events A *and* B. This is often called the *intersection* of the two events, or the *product* of the two events. The reason for these terms will be clear soon. For the moment we shall simple say {A *and* B}.

Suppose we throw two identical dice which are both fair, and are far from each other so that you expect that the result on one dice will not affect the probability of a result on the other.

Define the two events:

$$A = \{First\ dice\ shows\ 4\}$$

$$B = \{Second\ dice\ shows\ 6\}$$

We are interested in the probability of the event

$$\{A\ and\ B\} = \{First\ showed\ 4\ and\ second\ showed\ 6\}$$

A plausible reasoning to obtain the probability of these events is as follows: the specific result {*A and B*} is an *ordered* pair of numbers (the order is important – the number on the left is the outcome of the *first* dice, and the number on the right is the outcome on the *second* dice). We write the required event as {4,6}.

There are altogether 36 different pairs which are shown in the table below.

Table: All the 36 possible results for two dice

(1,1),(1,2),(1,3),(1,4),(1,5),(1,6)

(2,1),(2,2),(2,3),(2,4),(2,5),(2,6)

(3,1),(3,2),(3,3),(3,4),(3,5),(3,6)

(4,1),(4,2),(4,3),(4,4),(4,5),(4,6)

(5,1),(5,2),(5,3),(5,4),(5,5),(5,6)

(6,1),(6,2),(6,3),(6,4),(6,5),(6,6)

We assume that the two dice are fair, and that all the outcomes on the two dice are independent. Therefore, we argue that each event in the table has the same probability, i.e. one out of 36. Therefore, the probability of the required event is $P(4,6) = \frac{1}{36}$. Here, we use the double brackets to emphasize that the event is: "4" on the first on "6" on the second.

We know that the probability of obtaining a "4" on the first dice is 1/6. We also know that the probability of obtaining a "6" on the second dice is 1/6. We saw that the probability of the required event is

$$P(4,6) = \frac{1}{36} = \frac{1}{6} \times \frac{1}{6} = P(4) \times P(6)$$

The last formula is suggestive. Before we accept it as a general formula for any two independent events, let us try one more example of two independent events. We throw a fair dice and a fair coin simultaneously. We assume that the sample space of the dice has six outcomes

$$\{1,2,3,4,5,6\}$$

and that the sample space of the coin has two outcomes

$$\{H,T\}$$

What is the probability of the event: "outcome 4" on the dice and the outcome H on the coin?" The sample space of the joint experiment is written in the table below:

$$(1,H), (2,H), (3,H), (4,H), (5,H), (6,H)$$

$$(1,T), (2,T), (3,T), (4,T), (5,T), (6,T)$$

Altogether, there are 12 events, and if both the dice and the coin are fair, and that they were thrown at different locations so that one throw does not "know" about the other, it is reasonable to assume that all of these 12 outcomes are equally probable, and each has probability 1/12. Thus, we have

$$P(4, H) = \frac{1}{12} = \frac{1}{6} \times \frac{1}{2} = P(4) \times P(H)$$

Again, we see that the probability of the event "4" *and* "T" is the product of probabilities of the separate events, "4" and "T."

Having these two examples in mind we generalize to any two independent events. First, we denote by $A \cap B$ the event "A *and* B." Sometimes, because of the product rule above, one uses the notation $A \cdot B$ for the event $A \cap B$. This notation might be confusing because we did not define a "product" of the two *events*. The product of the two probabilities in the above formula is a product between two numbers, whereas the notation $A \cdot B$ is not a product between two numbers. Therefore, we shall use either the notation A *and* B, or $A \cap B$.

The general rule for any two independent events is:

$$P(A \cap B) = P(A) \times P(B)$$

This rule is valid provided we know that the two events are independent. In most cases we do not *know* whether or not the events are independent, therefore the above formula is taken as a *definition* of independence. Two events are said to be independent, if and only if, the probability of the event

$(A \cap B)$, i.e. (A *and* B) is the product of the probability of A and the probability of B.

Note carefully that in the two examples we worked up above, the events were elementary outcomes; a throw of a dice, or a toss of a coin. The rule above applies to any two independent events A and B, where A and B could also be compound events.

Note also that the formula above not only *defines* independence between two events, it also gives us the rule on how to calculate the probability of the event $(A \cap B)$, i.e. that *both* A *and* B occurred.

It is important to distinguish between the notions of *disjoint* events and *independent* events. We say that two events are disjoint when they do not have any elementary event in common. Examples are:

$$A = \{even\} \text{ and } B = \{odd\}$$

i.e. an "even" result on one dice and an "odd" result on the *same* dice. Clearly, if A occurs it means that B did not occur and vice versa. In general, two events are said to be disjoint when the occurrence of one of them excludes the possibility of the occurrence of the other.

Equivalently, we write for two disjoint events: $\qquad A \cap B = \emptyset$

where $\emptyset$ is the empty event, i.e. the impossible event. Thus, two events A and B are disjoint if the intersection between the two is empty. Therefore, the probability of the occurrence of $(A \cap B)$ i.e. (A *and* B) is zero.

Note however, that the event "even" on one dice and the event "odd" on a second dice are not necessarily disjoint. Both events can occur, and if they occur on two fair dice which are far apart, the probability would be simply $\frac{1}{2} \times \frac{1}{2} = \frac{1}{4}$.

The notion of *disjoint* is defined in terms of the *events themselves*. For instance, the following two events for the same dice are *not* disjoint:

$$A = \{1,2,3\}$$

$$B = \{3,4,5,6\}$$

On the other hand, the two events

$$C = \{1,2\} \quad , \quad D = \{5,6\}$$

on the same dice are disjoint.

Similarly, the event: "hitting the region A" and "hitting the region B" in Figure 4, are not disjoint events. On the other hand, "hitting the regions C" and "hitting the region D" are disjoint events.

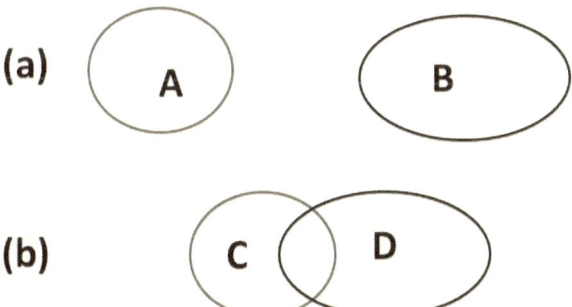

Figure 4. (a) Two disjoint events A and B, and (b) Two overlapping events C and D

In all the examples above about *disjoint* events, we did not mention the probabilities of any of the events. This notion is a property of the events themselves and not of their probabilities. On the other hand, the notion of *dependence* and *independence* are *defined* in terms of the probabilities of the involved events.

When two events A and B are *not* disjoint it means that they have some elementary events in common. For instance, the two events (on the same dice

$$\{1,2\} \quad \text{and} \quad \{4,5\}$$

are disjoint. On the other hand, the following events (on the same dice): {1, 2, 3} and {3, 4, 5} are not disjoint. They have one elementary event in common; {3}

The following events (on the same dice) have more elementary events in common

$$\{1,2,3,4\} \quad \text{and} \quad \{3,4,5,6\}$$

(5) Conditional probability

The concept of *conditional probability* is central to the theory of probability, perhaps even more central than the concept of probability itself. You will see why in a moment. But before defining the concept of conditional probability let us warm up with some simple examples. This will lead us not only to the motivation for defining the concept of conditional probability, but it will also point the way towards the formal definition of this concept.

Define the event A as follows:

A = {a dice shows the face with 4 dots upwards}

Answer the following questions:

1. What is the probability of the event A?

2. What is the probability of the event A given that the dice is fair?

3. What is the probability of the event A given that the dice is fair, and that we threw it in such a way that it rolled and spun in the air many times before it landed on the floor?

4. What is the probability of the event A if you know that the dice is tied to very short spring band, or a spring that pulls the rubber so that the one dot will always face downwards on one end, and that the second end is connected to the floor?

5. What is the probability of the event A given that a heavy metal was attached to the face having six dots (the one dot face is always parallel to the six dotted face).

The morale of this exercise is simple. The probability of any event is always a *conditional* probability. The condition is the given information. In most cases the only information that we give is that the experiment is performed. We can write it symbolically as $P(A|\Omega)$, i.e. the probability that the event A occurs given that a well-defined experiment is performed, and that one of the outcomes has occurred. We usually omit the condition "Ω" when we ask about the probability of an event A. However, one should be aware of the fact that there exists no such thing as an *absolute* probability of an event, without giving *any* additional information.

(6) The general definition of condition of probability

Now that we know that a probability is always a conditional probability, and that we already know the standard condition (i.e. that the die is fair, and the experiment is performed, etc.) we shall always omit this information from our notation, and refer to P(A) as the probability of the event A, given the standard information implicitly. The next step is to calculate the conditional probability of an event A given some *new* (non-standard) information.

I threw a dice and I tell you that the result was "even." Now, I ask you, "what the conditional probability that the outcome is "4", given that the result is "even?" We write this as

$$P("4"|"even")$$

where the condition is written on the right hand side of the vertical bar. (Note again the omitted standard information).

How do we calculate this conditional probability? The answer is very simple. The sample space for the dice is $\Omega = \{1,2,3,4,5,6\}$, and the probability of each outcome is 1/6. We write this as $P("4") = \frac{1}{6}$. Having the additional information "even," effectively *reduced* the sample space from Ω to the new sample space $\{2,4,6\}$. Each of these outcomes in the reduced sample space has probability 1/3, i.e. $P("4"|"even") = \frac{1}{3}$.

Now we also notice that this result can be obtained by the ratio

(a) $\qquad P("4"|"even") = \dfrac{P("4")}{P("even")} = \dfrac{\frac{1}{6}}{\frac{1}{2}} = \dfrac{1}{3}$

This method of calculation of the conditional probability is only suggestive. Whenever we give additional information we reduce the sample space from which we select the required event. But what if we are told that the result was "odd" and ask for the probability that it is "4" given that the result is "odd." Clearly, in this case the answer is zero. But if we write the ratio as we did above we shall get the incorrect result

(b) $$P(\text{"4"}|\text{"}odd\text{"}) = \frac{P(\text{"4"})}{P(\text{"}odd\text{"})} = \frac{1/6}{1/2} = \frac{1}{3}$$

This absurd result is again only indicative. The correct procedure to calculate the conditional probability is

(c) $$P(\text{"4"}|\text{"}even\text{"}) = \frac{P(\text{"4" }and\text{ "even"})}{P(\text{ "even"})} = \frac{1/6}{1/2} = \frac{1}{3}$$

(d) $$P(\text{"4"}|\text{"}odd\text{"}) = \frac{P(\text{"4" }and\text{ "odd"})}{P(\text{ "odd"})} = \frac{0}{1/2} = 0$$

where in the numerator in the first quotient (c), is $P(\text{"4" }and\text{ "even"})$ is simply 1/6. The numerator in (d) is $P(\text{"4" }and\text{ "odd"})$ is obviously zero. The outcome cannot be "4" and "odd" at the same time.

Before we announce the general definition of conditional probability let us work out another example which will provide us with an additional rationale for the general definition. It is also a good example for training you to think probabilistically.

We throw two fair dice at large distance from each other so that the outcomes on each of the dice are independent. What is the probability that *both of the outcomes* are "even," given that the *sum* of the outcomes is 8?

The sample space of this experiment contains 36 elementary outcomes.

Here they are:

(1, 1), (1, 2), (1, 3), (1, 4), (1, 5), (1, 6)

(2, 1), (2, 2), (2, 3), (2, 4), (2, 5), (2, 6)

(3, 1), (3, 2), (3, 3), (3, 4), (3, 5), (3, 6)

(4, 1), (4, 2), (4, 3), (4, 4), (4, 5), (4, 6)

(5, 1), (5, 2), (5, 3), (5, 4), (5, 5), (5, 6)

(6, 1), (6, 2), (6, 3), (6, 4), (6, 5), (6, 6)

Note that the result (1, 6) is different from the result (6, 1). The two dice are different and distinguishable. The first number in the parenthesis is the outcome on the first dice, and the second number is the outcome on the second dice.

Assuming that all these 36 elementary outcomes are equally likely, each has the probability of 1/36.

By inspection of the table we can write the following probabilities

(i) $$P(sum = 8) = \frac{5}{36}$$

There are five outcomes with sum = 8 out of 36 total outcomes. Therefore, based on the *classical* definition we arrive at the probability 5/36.

(ii) $P(sum = 8 \text{ and both outcomes are "even"}) = \frac{3}{36}$

There are only three pairs which include two even numbers *and* sum = 8. Finally, the required conditional probability is:

(iii) $P(both\ numbers\ are\ even|sum = 8) = \frac{3}{5}$

Given the condition $sum = 8$ reduce our sample space from 36 outcomes to only five outcomes Out of these five, only three outcomes consist of two even numbers.

Note that we derived the last result by applying the classical "definition" of probability in the reduced sample space. We did not use any general formula. Let us see what we get if we apply methods (c) and (d) to calculate the conditional probability. Applying method (c) we get,

(d) $P(both\ even|sum = 8) = \dfrac{P(both\ even)}{P(sum=8)} = \dfrac{9/36}{5/36} = \dfrac{9}{5}$

Note that there are nine outcomes with both entries "even," and nine outcomes with both entries "odd," and 18 outcomes with one "odd" and one "even" entry. Obviously, this method gave the wrong answer (in fact, the number is not even a probability, it is larger than unity). Let us try the second method (d).

(e) $P(both\ even|sum = 8) = \dfrac{P(both\ even\ and\ sum=8)}{P(sum=8)} = \dfrac{3/36}{5/36} = \dfrac{3}{5}$

Using this method we get the correct answer as the one we calculated by the "classical definition." Being encouraged by calculating the conditional probability using method (d) we can generalize the definition, as well as the method of calculating the *conditional probability* of event A, *given that* event B occurred as:

$$P(A|B) = \dfrac{P(A\ and\ B)}{P(B)}$$

This is a very important formula. We have obtained it by examining a few simple examples (I encourage you to invent other examples, and see that this formula gives the correct result. If you cannot invent an example, let me suggest one: Calculate the probability that the two outcomes on the two dice are "even" given that the sum of the outcomes is "4").

Keep in mind that there is always the hidden condition Ω that we omit for simplicity of the notation. Also, keep in mind that the formula given above is meaningful only if B is *not* the impossible event, i.e. that $P(B) \neq 0$. It is meaningless to say that an event which has a zero probability has occurred.

Note 2. Bayes' theorem

Bayes' theorem is very simple and is very useful in solving some probability problems, the answers to which are sometimes counter-intuitive. If you have absorbed the concept of conditional probability you should have no difficulty in understanding Bayes' theorem. It is nothing more than a small variation on the theme of conditional probability. Yet, this theorem is extremely useful in many applications. Unfortunately, it is sometimes misused and sometimes abused. Understanding this theorem is very satisfying and rewarding.

Let us warm up with a simple example. I throw a dart on a board A. The board is divided into two regions A_1 and A_2, such that the union of the two regions is equal to the entire region of the board, write this as $A = A_1 \cup A_2$. Figure 1.

I tell you that the dart hit the board, and that all the points on the board are equivalent. As always in this book, we shall not be interested in the exact mathematical *point* at which the dart hit the board. There is an infinite number of points. Instead, we shall assume that the board is divided into very small squares, as small as you wish, but the total number of these points is finite. Also, we neglect the possibility that the dart hit the line dividing between the two regions A_1 and A_2. Figure1.

What is the probability that the dart hit the region A_1?

Next, we ask what the conditional probability of hitting the painted area B is, *given* that the dart hit A_1? Given that A_1 occurred, the probability of hitting the painted area is simply the fraction of A_1 which is painted.

$$P(B|A_1) = \frac{1}{10}$$

Again, you can derive this result from the classical definition of probability. The total area of A_1 is a_1, and the painted area in A_1 is $\frac{10}{100}a_1$, therefore the fraction of the painted area in A_1 is $\frac{10}{100} = \frac{1}{10}$.

Likewise, we have the probability of hitting the painted area in A_2

$$P(B|A_2) = \frac{2}{10}$$

At this point, we have all the required probabilities:

(1) $P(A_1) = \frac{a_1}{a}, \quad P(A_2) = \frac{a_2}{a}$

(2) $P(B|A_1) = \frac{1}{10}, \quad P(B|A_2) = \frac{2}{10}$

These are sometimes referred to as the *a priori* probabilities, i.e. these are given in advance, and later we shall calculate *a posteriori* probabilities.

The Bayes' theorem is concerned with calculating a kind of "inverse" probabilities. You are given the probabilities as in (1) and (2), and you are asked about the *inverse conditional* probabilities of i.e. Given that the dart hit the painted area B, what is the probability that it is in area A_1? In other words, we want to find the conditional probabilities $P(A_1|B)$ and $P(A_2|B)$. You can see that in the latter probabilities the roles of A_1 and B (and of A_2 and B) are "reversed" compared with the role in (2). In other words we reverse the roles of the *condition* and the outcome events.

To calculate $P(A_1|B)$ we use the definition of the conditional probability (see session 4).

$$P(A_1|B) = \frac{P(A_1 \cap B)}{P(B)}$$

Remember $P(A_1|B)$ is the probability of A_1 given B. $P(A_1 \cap B)$ is the probability that A_1 *and* B occurred.

Next, we want to rewrite the right-hand side of the equation in terms of the quantities we already know. Using again the definition of the conditional probability, we have:

$$P(A_1 \cap B) = P(A_1)P(B|A_1) = \frac{a_1}{a} \times \frac{1}{10} = \frac{1}{40}$$

Note carefully that the quantity $P(B|A_1)$ is the probability of B *given* A_1. Furthermore, we already calculated the probability of hitting B, which we can rewrite as:

$$P(B) = P(B \cap A_1) + P(B \cap A_2)$$

$$= P(B|A_1)P(A_1) + P(B|A_2)P(A_2)$$

$$= \frac{1}{10}\frac{a_1}{a} + \frac{2}{10}\frac{a_2}{a} = \frac{1}{10} \times \frac{1}{4} + \frac{2}{10} \times \frac{3}{4} = \frac{7}{40}$$

The first line of the equation means that the region B is simply the sum of the areas that B "cuts out of A_1" and the area that B "cuts out of A_2." See the two areas B_1 and B_2 in Figure 1. This equality is sometimes referred to as the *total probability theorem*. As you can see, this theorem is nothing but a statement that the area of B in Figure 1, is the sum of the areas B_1 and B_2. Next, we wrote each of the joint probabilities in terms of the conditional probabilities and in the final step we plug-in the numbers we already have. The result we need is therefore

$$P(A_1|B) = \frac{\frac{1}{40}}{\frac{7}{40}} = \frac{1}{7}$$

Now, let us write the theorem in general terms.

Given n events, $A_1, A_2, \ldots, A_n$. All are *disjoint* events which means that for each pair of events A_i and A_j, their *intersection* is empty.

$$A_i \cap A_j = \emptyset, \quad (i \neq j)$$

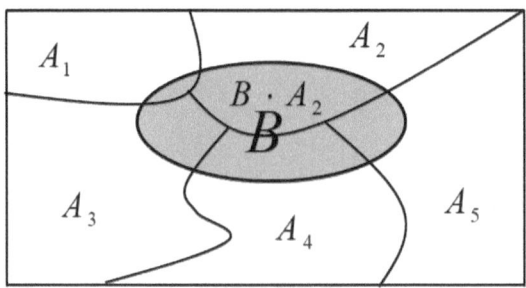

Figure 5. A region B intersecting the five regions A_i

See for example Figure 5. We also denote by A the *union* of all the events $A = \bigcup_{i=1}^{n} A_i$. Let B be any event (not the empty event), i.e. $B \neq \emptyset$ which intersects with all or part of the events $A_1, ..., A_n$. You are given the *a priori* probabilities

(1) $P(A_i) = p_i$

(2) $P(B|A_i) = q_i$

p_i is the probability that the event A_i occurred. q_i is the conditional probability that B will occur if it is known that A_i has occurred. In Figure 5 we show the entire region B in grey. In this particular example, we see that B does not intersect A_1, while B_3 is the intersection of B with A_3, i.e., some of the A_i have non-empty intersection with B, and some do not.

Now you are asked to write down the probability that A_i occurred when you know that B has occurred. The answer is straightforward. We start from the definition of the conditional probability

(3) $P(A_i|B) = \frac{Pr(A_i \cap B)}{Pr(B)}$

We rewrite the numerator using the definition of conditional probability but in reversing the roles of A_i and B:

(4) $$P(A_i \cap B) = P(B|A_i)P(A_i)$$

Can you guess why we did this?

Then, we rewrite the denominator of (4) in a more complicated way.

(5) $$P(B) = \sum_{i=1}^{n} P(B \cap A_i) = \sum_{i=1}^{n} P(B|A_i)P(A_i)$$

This simply means that if B intersects with some or all of the events A_i, then the probability of B is the sum of the probabilities of all the intersections $B \cap A_i$. This is the theorem of total probability.

Remember that we assumed that all the A_i are disjoint events. This means that the probability of the union of all A_i (denoted A) is the *sum* of the probabilities of each of the A_i. Convince yourself that if all A_i are disjoint events, then all the intersection events $B \cap A_i$ are also disjoint events.

Now rewriting (3) with the help of (4) and (5), we get the famous Bayes' theorem.

(6) $$P(A_i|B) = \frac{P(B|A_i)P(A_i)}{\sum_{i=1}^{n} P(B|A_i)P(A_i)}$$

This might look complicated, but as (I hope) you have convinced yourself, by working on the particular examples that this theorem is nothing but rewriting a conditional probability in a more complicated – albeit much more useful form.

Why is this so useful? Here, you have to look carefully at the last formula and see that on the right hand side of the equation there are all the quantities we *know*. On the left hand side is a quantity we want to calculate. The known quantities on the right hand side are sometimes referred to as *a priori* probabilities (given beforeh…and). The quantity on the left side is referred to as *a posteriori* (given after) probability. The reason is that in the *a priori* we go from A_i to B. In the *a posteriori* case we go in the "reverse" direction from B to A_i.

Note 3:   See Wikipedia GIGO

Note 4:

$$\alpha + \alpha = 2\alpha$$

$$\alpha - \alpha = 0$$

$$\alpha \times \alpha = \alpha^2$$

$$\alpha/\alpha = 1 \quad \text{unless } \alpha = 0$$

Note that α is an arbitrary number including zero.

Note 5: A proof, a la Unwin, that the probability of Kukuriku is 95%:

Here is the proof:

First, assume that the *a priori* probability of Kukuriku is 0.5. A fair assumption (according to Unwin). Next, we re-appraise this probability. The first "evidentiary area" is that we know that many children utter the sound "Kukuriku." Therefore, this information *supports* our assertion. We look at the "Divine" Table, and calculate that the new re-appraised probability is now 0.8. Second, in Israel children sometimes want to insult their friend by calling them Kuku. This is unrelated to the existence of Kukuriku, therefore its existence is unaffected by this "evidentiary area." The third "evidentiary area" is that we sometimes hear at daybreak, perhaps twice or even thrice, a sound emanating from the farm which reminds us of the existence of "Kukuriku." This certainly supports our assertion that Kukuriku exists. Now, our re-appraised probability is 0.95, or 95% – almost certainty!

You see, I have just proved that Kukuriku exists by 0.95 or 95%. I hope you were convinced.

You might think this is a joke. No, it is not a joke, not even a bad joke. It is sheer nonsense, disguised in a scientific language.

Note 6: The following note is for those who have no idea what a proof means, and therefore where misled by the author's usage of the term "proof".

Proof that $\sqrt{2}$ is not a rational number.

An irrational number is defined as a number that cannot be expressed as a ratio m/n of two integers $(n \neq 0)$.

Assume that $\sqrt{2}$ is a rational number we can write

$$\sqrt{2} = \frac{m}{n}$$

Without loss of generality we can assume that m and n have no common factor. If they have we can divide both numerator and denominator by that factor. For instance, if $m = 3$ and $n = 6$, we can divide by three and rewrite $\frac{3}{6}$ as $\frac{1}{2}$.

Take the square of this equation to get

$$2 = \frac{m^2}{n^2}$$

or $2n^2 = m^2$. This means that 2 is a factor of m^2. It follows that 2 is also a factor of m itself (if 2 did not divide m, then the factorization of m into prime numbers would not contain 2, and therefore also the square of m would not be divided by 2).

Therefore, we can write m as $m = 2x$, where x is an integer. Squaring the last equality gives $m^2 = 4x^2$ which means that $2n^2 = 4x^2$, or $n^2 = 2x^2$. Therefore, 2 divides n^2, and also must divide n.

We see that both m and n have a common factor (2) in contradiction to our assumption. Therefore, $\sqrt{2}$ cannot be a rational number.

References and suggested reading

Alexander, E., (2012), *Proof of Heaven. A Neurosurgeon's Journey into the Afterlife*, Simon and Schuster, Inc. New York.

Atkins, P. (2007), *Four Laws that Drive the Universe*, Oxford University Press.

Ben-Naim, A. (2008), *A Farewell to Entropy: Statistical Thermodynamics Based on Information*. World Scientific, Singapore.

Ben-Naim, A. (2010), *Discover Entropy and the Second Law of Thermodynamics. A Playful Way of Discovering a Law of Nature*. World Scientific, Singapore.

Ben-Naim, A. (2012), *Entropy and the Second Law. Interpretation and Misss-Interpretationsss*. World Scientific, Singapore.

Ben-Naim, A. (2015a), *Information, Entropy, Life and the Universe. What we know and what we do not know*. World Scientific, Singapore.

Ben-Naim, A. (2015b), *Discover Probability. How to Use It, how to Avoid Misusing It, and How It Affects Every Aspect of Your Life*. World Scientific, Singapore.

Ben-Naim, A. (2016a), *The Briefest History of Time*. World Scientific, Singapore.

Ben-Naim, A. (2016b), *Entropy the Truth the Whole Truth and Nothing but the Truth*, World Scientific Publishing, Singapore

Ben-Naim, A. (2017a), *Modern Thermodynamics*, World Scientific Publishing, Singapore.

Ben-Naim, A. (2017b), *Information Theory*, World Scientific Publishing, Singapore

Ben-Naim, A. (2017c), *The Four Laws that do not drive the Universe*. World Scientific Publishing, Singapore.

Ben-Naim, A. (2017d), *Entropy, Shannon's Measure of Information and Boltzmann's H-Theorem, in Entropy*, 19, 48-66, (2017d)

Brillouin, L. (1962), *Science and Information Theory*. Academy Press, New York.

Callen, H.B. (1985), *Thermodynamics and an Introduction to Thermostatics*. 2nd edition. Wiley, New York

Carroll, S. (2010), From Eternity to Here.The Quest for the Ultimate Theory of Time, Penguin Books, London.

Carroll, S. (2016), The Bog Picture. On the Origins of Life, Meaning, and the Universe Itself, Dutton, New York.

Cover, T. M. and Thomas, J. A. (1991), *Elements of Information Theory*. John Wiley and Sons,
New York.

Gamow, G. (1940), *Mr. Tompkins in Wonderland*. Cambridge University Press, Cambridge.

Gamow, G. and Stannard, R. (1999), *The New World of Mr. Tompkins*. Cambridge University Press, Cambridge.

Lloyd, S. (2006), *Programming The Universe, A Quantum Computer Scientist Takes On The Cosmos*, Alfred A Knopf, New York

Seife, C. (2006), *Decoding the Universe. How the Science of Information is Explaining Everything in the Cosmos, From our Brains to Black Holes*, Penguin Book, USA

Shannon, C. E. (1948), *A Mathematical Theory of Communication.* Bell System Tech. J., 27.

Unwin, S.D. (2003), *The Probability of God. A simple calculation that proves the Ultimate Truth.* Three Rivers Press, New York

www.ingramcontent.com/pod-product-compliance
Lightning Source LLC
Chambersburg PA
CBHW030008190526
45157CB00014B/1170